How Charles Darwin might have looked as a modern graduate student just back from five years of field work. This picture is intended to fix in readers' minds that Darwin was at his most innovative at this age, and readers should remember that Darwin might now be denied admission to a good graduate school because of his deficiencies in languages and math.

Evolution For Naturalists

EVOLUTION FOR NATURALISTS:
The Simple Principles and Complex Reality

P. J. Darlington, Jr.
Museum of Comparative Zoology
Harvard University

A Wiley-Interscience Publication
JOHN WILEY & SONS
New York • Chichester • Brisbane • Toronto

Library of Congress Cataloging in Publication Data

Darlington, Philip Jackson, 1904–
    Evolution for naturalists.

    "A Wiley-Interscience publication."
    Includes bibliographies and index.
    1. Evolution.   I. Title.

QH366.2.D34   575   79-22897
ISBN 0-471-04783-X

Printed in the United States of America

10 9 8 7 6 5 4 3 2 1

**To Ernst Mayr
in friendship and admiration**

PREFACE

ABOUT THE BOOK

This book is about organic evolution—the evolution of life on earth. It is intended as a reasonably complete, readable description and assessment of the theory of evolution as it stands now.

I want to describe organic evolution in plain words rather than mathematically, relate it to other processes of directional change in ways that will (I hope) promote understanding of it, and at the same time reassess current theories of it so critically as to have some chance of correcting old errors and discovering new things about it. Both professional biologists and naturalists should find something of interest in a book like this, and I have tried to write it in terms intelligible to both. The book might therefore be used as a starting point for reorientation or self-education in evolution by persons whose minds are not closed on the subject. It is not designed as a text, but might be used as the core of a somewhat unconventional, open-ended course or seminar rather than a conventional package.

Biologists should begin the book with Chapter 1, but naturalists may do better to begin with Chapter 9, which traces our own evolutionary history from the beginning, and then turn back to Chapter 1.

To write for both professional biologists and nonprofessional naturalists is not easy. Dawkins, in *The Selfish Gene,* says, "I have assumed that the layman has no special knowledge, but I have not assumed that he is stupid. . . . I have worked hard to try to [present] subtle and complicated ideas in non-mathematical language, without losing their essence." This is what *I* have tried to do, and I have assumed also that serious naturalists will work—read and think—to understand evolution. They must, if they wish to make intelligent decisions about themselves and their environment. Several of my chapters are somewhat technical, because they have to be, but naturalists should be able to get the essential points without reading every word. The role that I ask good naturalists to play in the study of evolution, and in application of evolutionary principles to people and problems, is outlined in the summary of Chapter 10. I ask a lot!

What is a naturalist? One definition is: a person who *sees* nature. Good naturalists not only see and take pleasure in individual plants and animals but also see communities and environments. And serious naturalists try to understand what they see, which means (among other things) understanding evolution. Naturalists *walk*; no other way of getting around lets them see as much. And most naturalists are self-taught.

Darwin was a naturalist who saw the plants and animals around him (and collected beetles), and who was self-taught. I do not think he ever had a formal course in biology, although he sometimes attended lectures. And he learned from friends; he was "the man who walked with Henslow," a botanist friend. His informal autobiography, written for his family, is fascinating reading for naturalists now. The edition to read is the uncensored one, edited by his granddaughter Nora Barlow (London, Collins, 1958; New York, Norton, 1969).

Professional biologists often take an organism out of its environment or model parts of evolutionary processes without seeing wholes. By doing this they learn things that could not be learned in any other way, but they too often fail to see reality as naturalists do. Good, serious naturalists should not defer to professional biologists. The reality that naturalists see is at least as important as the parts that professionals see. Naturalists and professionals can both profit by exchanging views, with mutual respect.

This book cannot be documented like a research paper. The reading and reference lists at the end of each chapter are like naturalists' walks through a forest. Naturalists cannot look everywhere at once, but usually focus on special things—on trees, or fungi, or birds, or insects—with the option of turning aside to anything exciting they find unexpectedly. Call it planned walking with serendipity. And even within a special focus naturalists can see only a small part of what is in the whole forest, and what they see and talk about depends partly on chance, on what they happen to find. My reading and reference lists are a naturalist's walks through a forest of literature. They focus on special aspects of evolution, with the option of turning aside to anything I have found exciting; and the books and articles listed and talked about (in parenthetical comments) depend partly on chance, on what I have happened to find. This is the only kind of reading and reference list that can be fitted into a book like this, and it may be the best kind for naturalists and sometimes for biologists too. (The lists are described in more detail in Chapter 3.)

I shall, here, look at evolution from three points of view. The first is that of a naturalist concerned with what we see, or ought to see, around us.

The second is that of a professional academic biologist, awed by recent spectacular advances in the understanding of life at all levels from molecules to the biosphere as a whole, and wanting to keep up to date without losing a realistic sense of proportion. And the third is that of a human being, one of 4 billion of us, appalled by the problems we face, and wanting to put them into useful perspective without pontification or propaganda. These three points of view go together remarkably well, as they should if we see clearly, for they are all views of the same process.

Of these three points of view, that of the naturalist concerned with situations and processes at the Darwinian level—the visible level of individuals and populations—is, I think, incomparably the most important. Evolution at this level has made us what we are, and in making us has made also our appalling problems, which rise from population explosions and human selfishness, aggression, and unadaptedness to our new social environment, all of which are products of evolution. I shall put into this book as much as I can of what naturalists need to know about evolution to understand what they see in the world around them and in themselves. And I shall use what naturalists see to exemplify, test, and sometimes falsify evolutionists' generalizations.

About the "Synthetic Theory" of Evolution

The current theory of evolution is a synthesis. It puts together more than a 100 years of observations, inferences, and experiments first by naturalists and later by geneticists, mathematicians, and molecular biologists. It is an immense accomplishment, and it is, I think, essentially correct. It also has immense force, which rises not only from its explanatory power but also from the dogmatic way it is accepted and taught by most biologists. But the synthetic theory is neither complete nor entirely correct, and the dogmatic power of it is dangerous. It should be criticized, and I shall criticize it here, but without belittling it. Perhaps I should add that, if there is an "Establishment" of evolutionists, I am presumably a member of it. I am therefore in a position to say that there is no conscious conspiracy among us against new or unconventional ideas, but that our minds may be closed involuntarily against them. Rex Stout remarks in a nonscientific context (in *Too Many Clients*) that "one of the brain's most efficient departments is the one that turns possibilities into probabilities, and probabilities into facts." I see this process going on not only in single brains (including my own), but also in sets of brains, and among both proponents and opponents of evolution theories. This process makes us all more dogmatic than we ought to be and fosters bitter controversies, some of which are considered in Chapter 10.

ABOUT THE AUTHOR

Readers of a book like this should know something about the author of it.

I am a professional biologist, 75 years old, and retired. My formal education was completed in 1931, before the mathematical era of biology had well begun, and long before the explosive evolution of molecular biology. That I have been an observer of these events rather than a contributor to them may, or may not, be a disadvantage. As a biologist, I have been (and still am) first a field naturalist and then a systematist. My work has taken me into remote parts of the world, including five years in the tropics of northern South America, the West Indies, tropical Australia, and (during World War II) New Guinea and the Philippines. I know at firsthand many of the living products of evolution, from insects to vertebrates, as they occur in undisturbed parts of the real world. My systematic work has been mainly with carabid beetles—"ground beetles"—some of which are to be found almost everywhere, and which show geographic and evolutionary patterns exceptionally well. Some of the examples I shall give in this book are drawn from carabids, and will be new to most readers. I am also a professional biogeographer, if there is such a thing; the relation of geographic patterns to evolution is one of my oldest and strongest interests.

Although I am, among working biologists, an old man, I shall try to avoid what I call "the old man's syndrome." Some elderly biologists seem to think anything they say is worth saying because they say it. They are too often sadly mistaken! This tendency of elderly biologists to overvalue themselves perhaps accounts for their choice of pictures of Darwin. The picture of Darwin that most people know, that is usually copied in textbooks, and that hangs in the hall of the Royal Society in London (a society chiefly of elderly men) shows him as a white-bearded septuagenarian. But this is not the man who originated the ideas we know him by. This was done by the young man who sailed on the *Beagle* and who then devoted himself to organizing and explaining what he had seen. I have chosen this man—Darwin the young naturalist, comparable to a modern graduate student—for my frontispiece.

Actually, some of the ideas I am writing about now date not from my older but from my younger years. I still have a package of manuscript written 50 years ago, when I was working for my Ph.D., on "Evolution of Associations" (something like what we now call "group selection"). William Morton Wheeler, my professor, wanted me to revise and publish it, but I put it off (wisely, I think; if I had published it, I might have spent the rest of my life defending wrong conclusions) in favor of extended field-work, systematics, and evolutionary biogeography. But multi-level evolution and related ideas have underlain my thinking and writing through the whole of my career. I have recently published some of these ideas in short papers in *The Proceedings of the National Academy of Sciences,* but it is time I put them into more complete and more coherent form. I think this is the most useful thing I can do in retirement now. And, of course, I want to do it.

P. J. DARLINGTON JR.

Cambridge, Massachusetts
January 1980

ACKNOWLEDGMENTS

I have not asked any biologists to read whole chapters of the book, but several persons have read short items and commented usefully on them or have given me special information. They include William L. Brown Jr., Frank M. Carpenter, G. R. Coope, Kenneth W. Cooper, G. W. Cottrell, Sidney Darlington, M. T. Ghiselin, George D. Snell, and Edward O. Wilson.

The late Carl H. Lindroth deserves special mention. I am indebted to him not only for permission to use maps published in *Opuscula Entomologica*, but also for a lifetime of exchange of information and ideas about Carabidae. He will be greatly missed!

I am indebted also to the editors of the following periodicals, who have cleared the way for me to use parts of my own papers: *Ecological Monographs, Evolution, Proceedings of the American Philosophical Society, Proceedings of the National Academy of Sciences USA*, and *Transactions of the American Entomological Society*.

For the highly professional drawings and diagrams I am indebted to Lydia Vickers.

And my wife has given me encouragement, criticism when I asked for it, and practical help in many ways.

CONTENTS

Evolution For Naturalists

PART ONE

Introductory Material

Four myopic evolutionists looked at evolution. One said, "It is survival of the fittest." One said, "It is differential repro- duction." One said, "It is change in gene ratios." And one said, "It is a molecular process." They all saw something real, but each magnified what he saw, and none saw evolution as a whole.

Original, but suggested by the Indian folk tale of the blind men and elephant.

1

HISTORY

The history of evolutionary ideas can be traced only briefly here. However, the theory of evolution has itself evolved by a process of diversification, competition, and selective elimination of ideas, and its evolution introduces important evolutionary concepts. The history of evolutionary ideas before Darwin exemplifies the randomness of events that precede selection. Darwin exemplifies the role of one mind in comparing ideas, eliminating (or, as we say now, falsifying) unfit ones, including the Biblical ones he began with, and making a coherent set of the ideas that survive. And the history of evolutionary ideas since Darwin, although of necessity reduced to trite generalizations here, introduces an important pattern, in which evolution proceeds by successive stages, but with the stages not only succeeding each other but also being added together, forming a succession *and* a sum.

Before Darwin

Evolutionary ideas through twenty-four centuries, from the Greeks to Darwin, are traced by Osborn, although his book is a somewhat superficial introduction to the subject and should be supplemented by other works listed in my chapter reference list.

The Greeks did have some ideas of directional change, or at least of sequences in the appearance of increasingly complex forms of life, but (as Osborn says, page 350) it is difficult to estimate how much they anticipated modern evolutionary thought. They came closer to anticipating the fact of evolution than the mechansim of it. Aristotle, for example, thought that an internal perfecting force caused plants and animals to approach idealized "types." The use of preconceived models, whether conceptual or mathematical, still characterizes "typologists" (now an opprobrious term) among evolutionists, taxonomists, and others.

Of later pre-Darwinian evolutionists only samples can be noted here whose ideas are of special interest or who were important historically. The sample is small, but nevertheless it suggests generalizations that are important and, perhaps, partly new.

Erasmus Darwin. Erasmus Darwin (1731–1802) was Charles' grandfather, a noted physician, and what we would now call a naturalist. He asked (*Zoonomia,*1974 reprint, page 505):

> From thus meditating on the great similarity of the structure of the warm-blooded animals [comparative anatomy], and at the same time [on] the great changes they undergo both before and after their nativity [embryology and ontogeny]; and by considering in how minute a proportion of time many of the changes of [domestic] animals above described have been produced; would it be too bold to imagine, that in the great length of time, since the earth began to exist, perhaps millions of ages before the commencement of the history of mankind, would it be too bold to imagine, that all warm-blooded animals have arisen from one living filament, which THE GREAT FIRST CAUSE endued with animality, with the power of acquiring new parts attended with new propensities, directed by irritations, sensations, volitions, and associations; and thus possessing the faculty of continuing to improve by its own inherent activity, and of delivering down those improvements by generation to its posterity, world without end!

He then extended the idea of ancestral filaments to other animals and to plants, and asked (page 507):

> Or, as the earth and ocean were probably peopled with vegetable productions long before the existence of animals; and many families of these animals long before other families of them, shall we conjecture, that one and the same kind of living filaments is and has been the cause of all organic life?

And he stressed extinctions (as shown by fossils) in relation to evolution of new forms, and noted that David Hume had suggested that the earth itself and its surface had evolved. But he did not use the words "evolution" or "evolved," at least not in the modern sense, but referred to "gradual formation and improvement."

This was a clear statement of the idea of evolution, but it was marred or compromised in several ways. The items quoted were buried in about 1000 pages of prose and poetry that included much nonsense and some loose philosophy of the kind that grandson Charles disliked. Erasmus Darwin suggested no realistic mechanism of evolution. And he assumed a deistic background, which Charles abandoned.

Lamarck.   Jean   Baptiste . . . Chevalier   de   Lamarck   (1744–1829)   was   a
professional biologist (a professor at the Paris Museum) and an influential pre-Darwinian evolutionist. He emphasized adaptation to the environment as evidence of the fact of evolution; postulated geologic time
sufficient for evolution of living things and thus probably influenced the
geologist Lyell; proposed an evolutionary classification of plants and animals; and helped prepare the intellectual environment for Darwin, although Darwin himself said he got nothing directly from Lamarck's
work. Although Lamarck was right about the fact of adaptive evolution,
he offered no satisfactory mechanism of it. However, he apparently did
not especially champion the idea of "inheritance of acquired characters," with which his name has been somewhat unjustly linked; the idea
was conventional in his time, and he accepted it uncritically.

This too-brief account of Lamarck should be supplemented by reference to Burkhardt (1977).

Patrick Matthew.   Patrick Matthew (1790–1874) was not a recognized
biologist but a Scotch landowner with a practical interest in timber
trees. Nevertheless, in 1831, the year Darwin started on the *Beagle* and
28 years before publication of the *Origin,* Matthew not only inferred the
fact of evolution (although he did not use this word) but also explained
it as a process of variation and selection. He noted the "almost unlimited diversification" of individual plants and animals, which would provide material for a "natural process of selection." He said:

> The self-regulating adaptive disposition of organized life may, in part, be
> traced to the extreme fecundity of Nature, who, as before stated, has, in
> all the varieties of her offspring, a prolific power much beyond (in many
> cases a thousand fold) what is necessary to fill up the vacancies caused by
> senile decay. As the field of existence is limited and pre-occupied, it is
> only the hardier, more robust, better suited to circumstance individuals,
> who are able to struggle forward to maturity, these inhabiting only the sit-
> uations to which they have superior adaptation and greater power of oc-
> cupancy than any other kind; the weaker, less circumstance-suited, being
> prematurely destroyed. This principle is in constant action, it regulates the
> colour, the figure, the capacities, and instincts; those individuals of each
> species, whose colour and covering are best suited to concealment or pro-
> tection from enemies, or defense from vicissitude and inclemencies of cli-
> mate, whose figure is best accommodated to health, strength, defense,
> and support . . . *those* only come forward to maturity from the strict ordeal
> by which Nature tests their adaptation to her standards of perfection and
> fitness to continue their kind by reproduction.

This was a remarkably comprehensive statement of the theory of evolution by natural selection. Within its awkward phrasing it used such now-familiar words as "diversification," "natural . . . selection," "organized" life, "struggle" for survival, "adaptation," "fitness," and "reproduction." It was published in a book, *On Naval Timber and Arboriculture,* that escaped the notice of evolutionists until Matthew himself called attention to it in 1860, the year after publication of the *Origin.* The preceding facts and quotations are chiefly from Wells (1973).

Robert Chambers. Robert Chambers (1802–1871), whose sweeping concept of the unity of cosmic and organic evolution is quoted at the beginning of my Part Two, was the author of *Vestiges of the Natural History of Creation,* which was first published in 1844, eight years after Darwin returned from his voyage on the *Beagle* and fifteen years before publication of the *Origin.* This work, like Erasmus Darwin's, combined legitimate evidence and arguments for evolution with a lot of unrealistic details, and it offered no clear mechanism for evolution. But it did have impact. Its several editions sold more than twenty thousand copies in Britain alone in the fifteen years before publication of the *Origin*—and remember how small the reading public was in those days.

At the time it was published, this was a dangerous book to write—and particularly so for Chambers, a publisher whose list included religious works. He therefore published the book anonymously, and even had his wife copy the manuscript so that his handwriting would not betray him. And the book was violently attacked. Bosanquet said of it:

> . . .We willingly accord to it all the attraction of novelty and ingenuity, the fascination of beauty, the delights of theory. We readily attribute to it all the graces of the accomplished harlot. Her song is like the syren for its melody and attractive sweetness; she is clothed in scarlet, and every kind of fancy work of dress and ornament; her step is grace, and lightness and life; her laughter light, her every motion musical. But she is a foul and filthy thing, whose touch is taint; whose breath is contamination; whose looks, words, and thought, will turn the spring of purity to a pest, of truth to lies, of life to death, of love to loathing. Such is philosophy without the maiden gem of truth and singleness of purpose; divorced from the sacred and ennobling rule and discipline of faith. Without this, philosophy is a wanton and deformed adultress.
>
> Before giving an outline of the scheme and theory which is elaborated in the "Vestiges of Creation," and combating the evil tendency and intention of the work, we think it right to show the depth and strength of the poison to which we would provide an antidote; . . .

(Book reviews had flavor in the midnineteenth century! What would the editors of *Science* or *Nature* say to a review like this now?)

The importance of this book is not that it influenced Darwin. His ideas were well formed before he could have read it, and there is not much in it he could have used in any case. And it probably did not convince many other persons. But it did help clear the way. It was one of the things that helped pre-advertise the idea of evolution so effectively that the first edition of the *Origin* sold out on the day it was published. And it drew the first, violent criticism that might otherwise have fallen on Darwin.

Edward Blyth.   Edward Blyth (1810–1873) was an English naturalist who spent twenty-two years in India, as Curator of the Museum of the Royal Asiatic Society of Bengal. He was a careful observer, and his papers contained much detailed information about Indian natural history. But he was not a pre-Darwinian evolutionist; he was a contemporary rather than a predecessor of Darwin, and he rejected the idea of evolution, until Darwin convinced him of it.

Several persons recently, notably C. D. Darlington in England and Loren Eiseley in America, have belittled Darwin, saying or implying that he was a mediocre man, who deliberately stole his principal idea, lied about it, and got away with it. Eiseley has accused Darwin specifically of stealing the idea of evolution by natural selection from Blyth. The weakness of this accusation is that Blyth himself did not accept selection as an evolutionary force, but as a stabilizing force which opposed change, which might produce varieties, but not new species. In 1837 Blyth actually asked, "May not a large proportion of what are considered species have descended from a common parentage?" And he rejected the idea!

I like to compare Eiseley's accusation against Darwin to the case a prosecutor might present against a suspected horse thief. Eiseley, as prosecutor, with Darwin in the dock, first shows that Blyth had a pasture—his field of interest in natural history. Eiseley then shows that Darwin was in Blyth's pasture—that he must have read Blyth's papers, which were published in well-known English natural history journals. And beyond this Eiseley gives evidence that Darwin picked flowers in Blyth's pasture—Darwin apparently took not only factual details but also actual phrases from Blyth's papers without acknowledging them. (Darwin *was* careless in citing sources of details in the *Origin.*) But the accusation is horse stealing, not flower picking, and no evidence is given that Blyth ever had a horse. In fact, he had none—no theory of evolution. His pasture was empty, and I, as self-appointed judge, therefore dismiss the charge of horse theft against Darwin.

Blyth does have a place in history, as a source of information which Darwin used. And what has been written about him exemplifies the difficulty and complexity of tracing the origins of evolutionary ideas. But the belittling of Darwin should not be taken seriously. The *Origin* contains more than enough evidence to prove his extraordinary originality and power of sorting evidence and reaching correct conclusions. I think the belittlers of Darwin cannot have read the *Origin* with much understanding of what is in it!

Summary of pre-Darwinians. It should be reemphasized that the few pre-Darwinian evolutionists, or almost-evolutionists, considered here are only samples, selected because they had more than usual influence or because we know more than usual about them. Others are known; Zirkle (1941) lists a number who recognized some components of the process of evolution by natural selection but "failed to make the ultimate logical inference" required to make a complete theory; and there were probably still others who are not remembered at all. We cannot hope ever to have a complete roster of them, still less to understand all the roots of their ideas and all their influences on each other. But even the few samples do suggest some generalizations.

Their diversity is striking. They include a physician (Erasmus Darwin), a professional biologist (Lamarck), a naturalist-curator (Blyth), a practical landowner-tree-farmer (Matthew), and an editor-publisher (Chambers). And Charles Darwin, a gentleman of independent means, adds still another category. It is difficult to find characteristics common to all these persons. Some were (for their times) meticulous observers, but this was hardly true of Chambers. Perhaps they were all unusually inquisitive, doubters by nature, and rejecters of authority, but this too is a dubious generalization. Charles Darwin himself, although inquisitive, was hardly by nature a doubter or rejecter. I suspect that chance had a lot to do with it, and that the diverse pre-Darwinian evolutionists were persons who happened to get the idea of evolution by accident or for trivial reasons which we do not know, and who then (the human mind being what it is) pursued the idea stubbornly and more or less dogmatically. This is to say that the original ideas may have been almost as random as gene mutations are.

A safer generalization is that the pre-Darwinian evolutionists all inferred the fact of evolution, but that few of them had any conception of the mechanism of it. Of those considered here, only Patrick Matthew suggested overproduction, variation, and natural selection as the evolutionary process.

Another safe generalization is that the pre-Darwinians did not have much influence on each other. To return to the random-mutation analogy, their ideas were like recurrent mutations which were not viable in the intellectual environment of their time, which did not survive (except as fossils on paper), and which formed no sets—no coherent theories of evolution—fit to survive and evolve as wholes. It was Darwin who first put his ideas together into a set that did survive and evolve as a whole, in an intellectual environment that had itself evolved.

And a final generalization is that chance has much to do with our judgments of the pre-Darwinians. A modern writer who happens to select, say, Blyth for special attention can make his subject seem more important than he was, and if it happens that no modern writer has "rediscovered" some other almost-evolutionist, the latter's role may be underestimated. Of course these and other cases are open to differences of opinion!

In this connection it is intriguing to compare the present reputations of Patrick Matthew and Gregor Mendel. Matthew had no direct influence on the evolution of evolution theory; if his publications were noticed at all, their significance was not recognized by biologists at the time. But this can be said of Mendel too; his work was rediscovered too late to be directly influential. And, while Matthew inferred evolution, Mendel apparently did not; if he broke away from traditional beliefs enough to accept the idea of evolution, he did not say so. Matthew, like Mendel, was concerned with what he observed in the real world; both were practical botanists. And surely the concept of adaptive change of species as a result of "extreme fecundity," "diversification," and "selection in nature" presented by Matthew was destined to be no less fruitful, when it was rediscovered, than the fact of particulate heredity. Yet we celebrate Mendel's centenary and not, or not so widely, Matthew's. Does this reflect a modern bias of biologists in favor of details that are easy to count, as opposed to principles that are difficult to observe in complex natural environments? Is it because Darwin overshadowed and inadvertently diminished Matthew in a way that the rediscoverers of Mendelism did not overshadow Mendel? Or is the emphasis on Mendel an historical accident? Perhaps it is a combination of all these things. Professor Kenneth Cooper (personal communication) suggests that the emphasis on Mendel began as a strategem to reduce arguments about priority among the three rediscoverers, de Vries, Correns, and Tschermak. Emphasis on Mendel's firstness (in 1866) made who was next first (in 1900) less of a bone to be contended for. But why do we still emphasize Mendel more than Matthew?

Charles Darwin

Charles Robert Darwin (1809–1882) was an unpretentious English gentleman-naturalist, who received a gentleman's education (appropriate earlier schooling and tutoring and three years at Cambridge) but did not distinguish himself in it. He considered first medicine and then the church as professions, but rejected both. His real interests at school and the university were shooting (for sport) and natural history, especially collecting beetles. Soon after his education was completed he was given the chance, almost by accident, to go as unpaid "naturalist" on Her Majesty's Survey Ship *Beagle* on a five-year voyage around the world. He went (1831–1836, Darwin being 22 to 27 years old); he saw the products of evolution in unspoiled parts of the world; and he came home to spend much of the next twenty-odd years gathering and organizing evidence that plant and animal species evolve by a process of variation and natural selection. And he published *On the Origin of Species by Means of Natural Selection* in 1859, after Alfred Russel Wallace had reached the same conclusions Darwin had, and after their joint presentation of the theory to the Royal Society in 1858. He then devoted the rest of his life to elaborating his theory and applying it to special problems, including the descent of man.

Darwin's own writings were voluminous, and books and articles about him, with accounts of his life and work, continue endlessly. Some are critical to the point of accusing Darwin of intellectual theft, as previously noted under Blyth; but the accusations seem to me unjustified. Darwin did not live in a vacuum. He took facts from many naturalists, and he sometimes skimped acknowledgments, perhaps partly because he did not remember where he got some details and partly because he thought of the *Origin* as an "abstract" which should not be overloaded with citations. Accusations of deliberate dishonesty aimed at Darwin are against the evidence and against what we know of his character. Of recent books that consider Darwin and his work with more understanding, Ghiselin's *The Triumph of the Darwinian Method* is notable. Of Darwin's own works, the one above all that modern evolutionists ought to *read* is the *Origin,* the facsimile of the first edition. And to understand Darwin as a person the book to read is the informal autobiography that he wrote for his family, and the edition to read is the one edited by his granddaughter, Nora Barlow; all earlier editions are marred by family censorship.

It is hard not to hero-worship Darwin, not because of his fame, but because of the extraordinary qualities of his work, especially of the *Origin,* and of his mind. These qualities are worth illustrating.

First and beneath everything else, Darwin was a population-thinker. He did not use the word "population," at least not in the first edition of the *Origin,* but that is what he often meant when he said "species." This fact has been emphasized only recently. Darwin had to think in terms of individuals *and* populations to make a workable theory of evolution. It is noteworthy, too, that when he talked about the origin of species he was usually thinking, not of how one species splits into two, but of how one species or population evolves to become a different species.

Too much is sometimes made of the fact that Darwin did not understand the mechanism of variation. It is true that he did not, and that he made some bad guesses about it later. But in the first edition of the *Origin* (page 486) he said, "A grand and almost untrodden field of inquiry will be opened [in the future], on the causes and laws of variation . . ." And the field *was* opened by Mendelian geneticists almost half a century later, and reopened by molecular biologists after almost another half century. Of course Darwin did not foresee this in detail, but he did see the limits of his knowledge. He observed variation in both domestic and wild plants and animals, and saw that it was sufficient for his theory, but he knew that he did not understand it.

Darwin's biogeography may be a better measure of his capacity than his "discovery" of evolution is. I have described it elsewhere (1959) in an article written to be read, not just filed in historical archives. Something more will be said about it in Chapter 2. Now, only one small but revealing detail need be noted. Darwin considered continental drift, although not by name (the *Origin* was published 53 years before Wegener's presentation of the drift hypothesis). What Darwin did do was to ask whether there was evidence of "such vast changes in [the] position and extension [of the continents] as to have united them within the recent period"—"position" and "united" are the significant words. And he concluded that "as far [back] as we can see" the oceans and continents "have stood where they now stand." But then he added, "but long before [then] the world may have presented a wholly different aspect." The point is, not that Darwin was right or wrong, but that he first asked himself a question that few persons asked seriously for another 100 years, answered it as well as he could from the evidence he had, stopped where he thought the evidence stopped, and kept an open mind. Very few evolutionists or biogeographers, and very few people, can manage their minds as well as this. I do not use the word "genius" lightly, but I think that Darwin's ability to criticize his own ideas and keep an open mind was a manifestation of genius.

Another noteworthy point is that when Darwin was on the *Beagle* and turned his mind around from a literal belief in the Bible to his first idea of evolution, he was a young man, about the age of a modern graduate student. The old sage with the long beard, whose picture is more familiar, deserved the honors that he received, but we should look at the young man if we wish to understand scientific innovation. This exceedingly important fact is emphasized also in my Frontispiece and Preface.

Darwin was not conspicuously intellectual. His IQ, tentatively calculated long after his death (Cox, 1926), has been estimated at 135 in childhood and 140 in young manhood—good, but not extraordinary. In his autobiography Darwin himself says that he was an "ordinary" student, "singularly incapable of mastering any language," slow in mathematics ("I do not believe that I should ever have succeeded [in math] beyond a very low grade"), and handicapped by "my incapacity to draw ... an irremediable evil." Some modern educators, including the late Harvard President Conant (1956, page 45), have wanted to make ability in mathematics and languages the test of genius and to select students accordingly. By this test, Darwin might not be allowed a first-class college education today. This is *not* farfetched and *not* funny.

Nevertheless, to do what Darwin did required literally extra-ordinary capacities. What were they? He was not just lucky; it is true that the time had come for the "discovery" of evolution, but why was Darwin the one who not only discovered it but presented a full-scale theory of it, based partly on population thinking, presenting the evidence and also recognizing the limits of the evidence, and inferring from it all that he did infer, from explanations of geographic patterns to a tentative history of man? And Darwin was not a "late bloomer." Whatever his capacities were, he had them when he was a young man; his key ideas came to him when he was still in his twenties. Perhaps he should be called not a late but a cryptic bloomer, whose capacities were there and might have been detected by an acute observer, but were more general than quantifiable, difficult to see and measure, of a sort that older men did not then and still do not easily recognize in younger men.

What were these qualities, so hard to measure but so extraordinarily productive? Nora Barlow, his granddaughter, says of Darwin, "The love of close observation of natural fact and his need for a theory to explain everything he saw, forms the closely woven tissue which constituted his genius." This is true, but it is not an explanation. It is also true, but obvious and trite, that his genius involved exceptional powers of concentration, persistence, and analysis. And beyond this he had one ability, already noted, which is less obvious but essential: the power of self-criticism. He could do more than formulate theories. He could look at his own theories and see what was wrong with them, correct and improve them, and see what their limits were, and he could do this without inhibiting himself. This is to say that he was superbly able to use the hypothetico-deductive method, which depends not only on formulation of ideas but also on elimination of erroneous ones. To use this method as Darwin used it requires almost unlimited time, which he had, thanks to inherited money and perhaps thanks also to continual poor health, which was not disabling but permitted him (or which he used?) to limit most activities other than the reading, thinking, experimenting, and writing that he wanted to do.

History Since Darwin

Four stages can be recognized in the evolution of evolutionary theory since publication of the *Origin,* each stage characterized by new discoveries or new methods.

Nineteenth-Century Darwinism. The first stage lasted until about 1900. During it, evolutionists, including Darwin himself until his death in 1882, were primarily concerned with naturalists' observations of the visible world, and with scientific and philosophical theories, arguments, and applications of the concepts of evolution—especially competition and natural selection—that Darwin had formulated. The concepts and applications were stubbornly resisted by the clergy and by many other persons, and this resulted in bitter controversies which cannot be detailed here.

Mendelian Heredity and Evolution. The second stage spanned two or three decades beginning with the rediscovery of Mendelian heredity in 1900, and was characterized by the rise of Mendelian genetics, which was at first derided and stubbornly resisted by some of the best of the older biologists including, I am sorry to say, my old professor William Morton Wheeler. This was the beginning of what Darwin had predicted, the opening of a "grand and almost untrodden field of inquiry . . . on the causes and laws of variation." Mendelian genetics made the most significant advances in evolution theory during this time.

Evolutionary Mathematics. The third stage was characterized by the rise of evolutionary mathematics, beginning with the work of Volterra, Lotka, Haldane, Gause, and Fisher in the 1920s and 1930s. Fisher's (1930) *The Genetical Theory of Natural Selection* was especially influential. Application of mathematics to evolution theory has been stubbornly resisted too, by persons like myself, not for its legitimate use but for its frequent excesses and unrealisms.

Molecular Biology and Evolution. The fourth stage, of course, was dominated by the rise of molecular biology, which had its inconspicuous beginnings but became dominant with the Watson and Crick explanation of the structure and replication of DNA in 1953. Molecular biology has dramatically increased our understanding of the mechanism of heredity and variation of living organisms, and of the origin and evolution of life. If the application of molecular biology to evolution has been resisted, it is less for itself than for its tendency to crowd out other kinds of biology in many university departments and in some biological journals.

Summary of History Since Darwin. Something can be said about the relation of these four stages to reality. The first stage was concerned with observation and explanation of the real, visible world. The second and fourth stages were concerned with the real world too, but primarily with mechanisms of heredity and variation and only secondarily with evolution. While the third, mathematical stage was not *primarily* concerned with the real world but with making models, which sometimes did but too-often did not fit reality.

These four historical stages not only followed each other but also overlapped and were added together. They formed, not just a sequence, but a cumulative sequence or stepped whole. The activities that characterized them all continue, very actively. And they interact. Naturalists' observations continue; they have become in part modern numerical ecology, but nonnumerical observations of the kind that Darwin made are still essential to emphasize the complexity of the real world, to criticize evolutionary mathematics and keep it in perspective, and to provide background for understanding human problems. Mendelian genetics still continues, fruitfully, and is being applied more and more to real situations. And, it is hardly necessary to add, molecular biology is thriving, also fruitfully.

The Present

The present book is intended as a reasonably complete, readable, critical description and assessment of the theory of evolution as it now stands. Books that treat evolution in other ways are cited in the chapter reading and reference list.

The Future

If a fifth stage in the evolution of our understanding of evolution is to be characterized by fundamentally new discoveries or concepts, the discoveries will probably be made at submolecular or subatomic levels. The outstanding evolutionary mystery now is how matter has originated and evolved, why it has taken its present form in the universe and on the earth, and why it is capable of forming itself into complex living sets of molecules. This capability is inherent in matter as we know it, in its organization and energy. Matter does form itself into sets. Selection only determines which sets disintegrate and which (for a time) do not.

However, while we wait for further discoveries at levels below the atom, we may be entering something like a new stage in the evolution of evolution theory. It is, or will be, characterized by no single dramatic event but by a more even synthesis and reemphasis on simple principles and precise and consistent definitions, without forgetting details and complexities. Above all it will be characterized by a return to reality, by an insistence that models of evolution, including mathematical ones, must fit the *real* real world. This return to reality will have several secondary effects, some of which are already visible. Departments of biology will find a better balance between evolutionary and molecular biology; fieldwork—direct study of real situations in nature—will become a primary part of biological education, not just incidental to lectures and laboratories; and this will require more emphasis on systematics than is usual now. We may hope that new generations of graduate students will require these new emphases, whatever the older faculties may think. And scientific journals will change their emphases too, giving a larger share of space to the real world as opposed to isolated fragments and oversimplified mathematical models of it.

Summary

The theory of evolution has evolved by stages, in an environment which has itself evolved. Before Darwin, ideas about evolution were usually random, fragmentary, uncoordinated, sometimes unrealistic, and unconvincing. Darwin put together observations and ideas into a coordinated theory which was complete (at the visible level), realistic, and therefore convincing. After Darwin, naturalists, Mendelian geneticists, mathematicians, and molecular biologists have successively elaborated and illuminated the theory; their contributions form a multi-level "stepped whole," within which research continues fruitfully at all levels. The product of this evolutionary process is the present "synthetic theory" of evolution. In the future, fundamentally new discoveries are most likely to be made at submolecular levels, but in the meantime we seem to be entering a new, healthy stage of more-even synthesis and return to reality in the study of evolution.

READING AND REFERENCES

The following are selected fragments from the enormous literature on the history of evolution theory.

Before Darwin

Darwin, C. 1861. An Historical Sketch of the Recent Progress of Opinion on the Origin of Species. (In) *The Origin,* 3rd ed. (A short, imperfect account of Darwin's predecessors as he saw them.)

Osborn, H. F. 1929. *From the Greeks to Darwin,* 2nd ed. New York and London, Charles Scribner's Sons.

Zirkle, C. 1941. Natural Selection Before the Origin of Species. *Proc. Am. Philos. Soc.,* **84,** 71–123.

Erasmus Darwin

Darwin, E. 1794–1796. *Zoonomia; or The Laws of Organic Life.* (See the following book for references to this and later editions, which include a 1974 reprint, New York, AMS Press.)

King-Hele, D. 1977. *Doctor of Revolution. The Life and Genius of Erasmus Darwin.* London, Faber & Faber.

Lamarck

Burkhardt, R. W., Jr. 1977. *The Spirit of System. Lamarck and Evolutionary Biology.* Cambridge, Massachusetts, Harvard Univ. Press.

Patrick Matthew

Wells, K. D. 1973. The Historical Context of Natural Selection: The Case of Patrick Matthew. *J. His. Biol.,* **6,** 225–258.

Robert Chambers

Anonymous [Chambers, R.]. 1844. *Vestiges of the Natural History of Creation.* London, John Churchill. (Later editions include, 1969, a reprint of the first, with an introduction by de Beer, Leicester, England, Leicester Univ. Press; New York, Humanities Press.)

Bosanquet, S. R. 1845. *"Vestiges of the Natural History of Creation:" Its Arguments Examined and Exposed.* London, John Hatchard & Son.

Millhauser, M. 1959. *Just Before Darwin. Robert Chambers and Vestiges.* Middletown, Connecticut, Wesleyan Univ. Press.

Edward Blyth

Eiseley, L. C. 1959. Charles Darwin, Edward Blyth, and the Theory of Natural Selection. *Proc. Am. Philos. Soc.,* **103,** 94–158. [This paper has been reprinted in a book of Eiseley's essays, published since his death in 1977, and criticized by Gould (1979, see below under *Charles Darwin*).]

Darlington, C. D. 1959. *Darwin's Place in History*. Oxford, Basil Blackwell.

Beddall, B. G. 1973. Notes for Mr. Darwin: Letters to Charles Darwin from Edward Blyth at Calcutta: A Study in the Process of Discovery. *J. Hist. Biol.*, **6,** 69–95.

Charles Darwin

Publications about Darwin continue endlessly. They include occasional new ideas (which are still possible even about Darwin), much conscious and unconscious repetition, and some nit-picking. The whole has evolved into what can fairly be called a Darwin industry, profitable to scholars, popularizers, and publishers. The following works are, of necessity, selected somewhat arbitrarily.

Darwin, C. 1839. *Journal of Researches into the Geology and Natural History of the Various Countries Visited by H.M.S. Beagle*. London, Henry Colburn. (Other editions include a facsimile reprint of the first, 1952, New York and London, Hafner Publ. Co. For a shorter, readable account of the voyage *see* Moorehead, A., 1969, *Darwin and the Beagle*, New York, Harper & Row. And for previously unpublished pencil drawings and water colors by the Beagle's official artist, Conrad Martens, with appropriate text taken from letters of Darwin and Fitzroy and other sources, see Keynes, R. D., 1979, *The Beagle Record*, Cambridge England, and New York, Cambridge Univ. Press.)

Darwin, C. 1859. *On the Origin of Species by Means of Natural Selection, or the Preservation of Favored Races in the Struggle for Life*. London, John Murray. (Numerous later editions include a facsimile of the first with an introduction by Ernst Mayr, 1964, Cambridge, Massachusetts, Harvard Univ. Press; and a variorum text edited by Morse Peckham, 1959, Philadelphia, Pennsylvania, Univ. Pennsylvania Press.)

Darwin, C. 1958. *The Autobiography of Charles Darwin, 1809–1882, with Original Omissions Restored*, edited with appendix and notes by his granddaughter Nora Barlow. London, Collins.

Cox, C. 1926. *The Early Mental Traits of Three Hundred Geniuses*. Stanford, California, Stanford Univ. Press.

Conant, J. B. 1956. *The Citadel of Learning*. New Haven, Connecticut, Yale Univ. Press.

Various authors and editors. 1959 (±). Numerous books and articles marking the centenary of publication of the *Origin; see* library subject indices.

Darlington, P. J., Jr. 1959. Darwin and Zoogeography. *Proc. Am. Philos. Soc.*, **103,** 307–319. (What biogeography did for Darwin and what Darwin did for biogeography, with an attempt to define his genius.)

Ghiselin, M. T. 1969. *The Triumph of the Darwinian Method*. Berkeley and Los Angeles, California, Univ. California Press.

Gould, S. J. 1979. Darwin Vindicated! *New York Review of Books*, August 16, 1979, 36–38. (A well-informed discussion of the origins of Darwin's ideas.)

History Since Darwin

Nineteenth Century Darwinism.

The following are two samples from an extensive literature, selected because they were important and are still easy to obtain or to see in libraries. The first supports Darwin; the second criticizes him.

Huxley, T. H. 1893–1894. *Darwiniana.* (A volume of essays published in England from 1893 to 1894, with an American edition in New York in 1896, reprinted 1970, New York, AMS Press.)

Butler, S. 1886. *Luck or Cunning.* London, Fifield.

Mendelian Heredity and Evolution.

Dunn, L. C. 1965. *A Short History of Genetics.* New York, McGraw-Hill.

Stern, C., and E. R. Sherwood. 1966. *The Origin of Genetics. A Mendel Source Book.* San Francisco, Freeman.

Zirkle, C. 1968. The Role of Liberty Hyde Bailey and Hugo de Vries in the Rediscovery of Mendelism. *J. Hist. Biol.,* **1,** 205–218. (An account of the triple rediscovery of Mendelism in 1900 and the bickering that followed.)

Dobzhansky, T. 1967. Looking Back at Mendel's Discovery. *Science,* **156,** 1588–1589. (Review of six books on Mendel and the history of genetics.)

Evolutionary Mathematics.

Fisher, R. A. 1930. *The Genetical Theory of Natural Selection.* Oxford, Clarendon Press.

Molecular Biology and Evolution.

Watson, J. D., and F. H. C. Crick. 1953. A Structure for Deoxyribose Nucleic Acid. *Nature,* **171,** 737–738. (And) General Implications of the Structure of Deoxyribose Nucleic Acid. *Nature,* **171,** 964–967.

Watson, J. D. 1968. *The Double Helix,* New York, Atheneum. (A possibly prejudiced history of the discovery of the structure of DNA, reviewed and criticized many times. *See* Stent, G. S. 1968, What They Are Saying about Honest Jim, (a review of the reviews) *Quart. Rev. Biol.,* **43,** 179–184. And *see also* Sayre, A., 1975, *Rosalind Franklin and DNA,* New York, Norton, and the review of this book in *Quart. Rev. Biol.,* **51,** 285–289.)

Various authors. 1974. Twenty-one Years of the Double Helix. *Nature,* **248,** 721, 766–788. (A set of essays.)

Olby, R. 1974. *The Path to the Double Helix.* London, Macmillan. (*See* under the following book.)

Judson, H. F. 1979. *The Eighth Day of Creation. The Makers of the Revolution in Biology.* New York, Simon and Schuster. (A fascinating book; noticed in *Time;* reviewed in the *New York Times* book review section, April 8, 1979; and in part serialized in the *New Yorker.* But it is not easy reading; it is concerned with people more than biology; and its focus is outrageously narrow. Molecular discoveries are presented as if they are all that is worth knowing about modern biology and are scarcely related at all to the kind of evolution that has made the living world, including us and our problems. This distortion of reality is as great as if the molecular chemistry of gasoline were presented as all that is worth knowing about the modern automobile. Olby's book is simpler and better balanced and is fascinating too; Crick helped with it and wrote the foreword.)

The Present

Moore, R. 1962. *Evolution*. (In) Life Nature Library, New York, Time Inc. (Probably the best very short, easily available, conventional introduction to evolution.)

Patent, D. H. 1977. *Evolution Goes On Every Day*. New York, Holiday House. (A small book that presents modern aspects and implications of evolution in more detail than the title suggests.)

Dobzhansky, T., et al. 1977. *Evolution*. San Francisco, Freeman. (A longer, more technical, up-to-date, but still conventional, treatment of the synthetic theory.)

Various authors. 1978. *Evolution. Scientific American,* September (An authoritative, well-illustrated set of articles, with a superb reproduction of George Richmond's portrait of Darwin at age 31.)

The Future

I know no book that is specifically concerned with the future of evolution theory. More-general books include the following:

Cornish, E. 1977. *The Study of the Future*. World Future Soc., 4916 St. Elmo Ave. (Bethesda), Washington, D.C. 20014.

Robinson, T. C. L. 1977. *The Future of Science*. New York, Wiley.

Waddington, C. H. 1978. *The Man-made Future*. New York, St. Martin's Press. (Reviewed in *Quart. Rev. Biol.,* **54,** 199–200.)

2

EVIDENCE

Evidence that evolution has in fact occurred is one thing, evidence of how it has occurred, another, although an acceptable explanation of the how (Darwin's natural selection) was necessary to general acceptance of the fact. This chapter is concerned chiefly with the fact of evolution. Much of the rest of the book is concerned with the how.

Texts on evolution usually review the evidence that it has occurred, but too often present it as from authority. Here, emphasis will be on the evidence that convinced Darwin, and on similar evidence that naturalists can see now.

When Darwin sailed from England on the *Beagle,* he believed the Bible literally. What he saw during the voyage, as described in his *Journal* and later "abstracted" in the *Origin,* was therefore historically decisive, and is still decisive. In the first sentence of the *Origin* Darwin says:

> When on board H.M.S. Beagle, as naturalist,* I was much struck with certain facts in the distribution of the inhabitants of South America, and in the geological relations of the present to the past inhabitants of that continent. These facts seemed to me to throw some light on the origin of species—that mystery of mysteries . . .

* Recent commentators, including Gould (1977, pages 28–33) in an otherwise excellent essay on Darwin and Fitzroy (Captain of the Beagle), have argued that Darwin was not the naturalist on the *Beagle,* but that the surgeon, Robert McKormick, was. No official record supports this argument, but only a letter from a Scotchman addressed to McKormick as if he were to be the naturalist, and the simplest explanation is surely that the writer of the letter was mistaken. Against it we have Darwin's own statement that he was aboard the *Beagle* as the naturalist, and the well-known fact that Captain Fitzroy was looking for a working naturalist when he enlisted Darwin—who almost lost the place because Fitzroy doubted his persistence. Persons who question Darwin-as-naturalist protest too much that they are not nit-picking. But nit-picking is legitimate fun; why pretend that it is anything else?

This quotation makes clear that the evidence that first persuaded Darwin to think seriously about evolution came from biogeography—from the geographic patterns that evolving plants and animals make on the earth's surface, and from the geographic occurrence of fossils, in the same regions as their living relatives. I shall return to these subjects, but want first to summarize some other evidence.

Variation Under Domestication

The first chapter of the *Origin* is concerned with "variation under domestication." or with the evolutionlike modification and diversification of plants and animals under selection by man. Darwin stressed the many forms of domestic pigeons that have been derived from one wild ancestor, but now most persons will be more familiar with the many forms of dogs that have been derived from one (or a few) wolflike ancestor(s). These are obvious examples that anyone can see. Many other animals and many plants have been conspicuously modified by selection by man, and this is visible proof that species are capable of modification, although the changes produced by man differ in some ways—they are usually more rapid and less stable—from changes produced by evolution in nature. Indian corn or maize is a notable example, described by Mangelsdorf (1974).

Less direct evidences that evolution has occurred come from virtually every subfield of biology. Darwin noted them from "morphology" and embryology and from the fact that plants and animals can be classified into groups of "related" forms consistent with evolution from common ancestors. Many of the groups were recognized before acceptance of the idea of evolution and can fairly be considered evidence for it. And now we have evidence also from post-Darwinian genetics and physiology and most recently from molecular biology, which reveals the essential uniformity and continuity of all living things.

Comparative Anatomy

Forelimbs. Comparative anatomy—Darwin's "morphology"—is one of the sources of evidence that interested persons can see for themselves. For a trite but very good example, comparisons of the forelimb skeletons of a frog or lizard, squirrel or cat, horse, bird, bat, and man show detailed similarities that are difficult to account for except by derivation from a common ancestor. Even amateur naturalists can obtain some forelimb skeletons for themselves, and series of forelimbs are figured and compared in many texts and other works, for example by de Beer (1971).

The Human Hand and Foot.   The evolution of the hand is fascinating (see Napier, 1976), but the human foot tells a more complex story. Anyone can see that our five-toed foot is similar in a general way to other land-vertebrate hind feet (which are remarkably similar to fore feet in unspecialized quadrupeds), and the arrangement of bones is nearly the same. Our foot may therefore be supposed to have evolved from the common ancestor of four-footed land vertebrates. But, although our foot is now used principally for walking and running, it does not seem as well adapted for this use as the foot of a dog or a horse. This adaptive imperfection is well explained by our evolutionary history: the foot was apparently first adapted for running and claw climbing (as in tree shrews), then evolved to a handlike form adapted to grasping and hand climbing (as in monkeys), and is now evolving back to a walking-running form. It seems, therefore, that while our hand is a remarkably efficient product of evolution, our foot is less efficient, but is probably now evolving more rapidly. The last stages of its evolution are suggested by comparison with the feet of various apes (see Reynolds, 1967, pages 84–85, plate 32).

Embryology, Ontogeny, and Recapitulation

The ontogeny of an individual animal—its embryonic and later development—does obviously recapitulate its evolution, even though imprecisely, and ontogeny can be looked at as a programmed segment of evolution (see Chapter 5, "The biogenetic law"), but this is perhaps consistent with the fact of evolution rather than visible evidence of it.

The Fossil Record

The most nearly direct evidence of the evolution of particular groups of plants and animals is the fossil record, treated by Darwin in chapters IX and X of the *Origin*. The record is better now than when Darwin knew it. We now have important fossil "links" that were unknown to Darwin, including crossopterygian lobe-finned fishes with air-breathing lungs and walking fins appropriate to moving from water onto land, and famous *Archaeopteryx*, which links birds to their reptile ancestors (Chapter 8). We now have also sequences in the record showing (for example) the evolution of one-toed horses from four-toed ancestors, and, most exciting of all, of modern man from his prehuman ancestors (Chapter 9).

Many gaps and ambiguities occur in the fossil record and are stressed by critics, but (as Darwin noted) they are expected. Fossilization is and must be rare and chancy. If all the plants and animals living at one time were fossilized, they would fill the surface of the earth, and there would be no room for earlier or later forms. Fossil-bearing rocks actually cover only a small part of the earth's surface, and what is at the surface now, after being buried to fossil-forming depths and uncovered again, is an infinitesimally small sample of floras and faunas spanning more that three billion years of time. And the fossils themselves are usually only parts or fragments of individuals. The fossil record *must* be fragmentary. It would almost be more logical to criticize the record, not because it is incomplete, but because it is better than it ought to be. If this criticism were made, the answer would be that fossils do not occur at random, but chiefly in places where conditions for their formation at particular times were especially favorable, and where geological processes have happened to bring them to the surface. The record of some groups at some places and some times is therefore relatively good. The limits of the fossil record are further considered in Chapter 8.

Geographic Patterns

South America and the Galápagos.   To come back to what first turned Darwin's attention to evolution: in South America and the Galápagos he saw three related geographic patterns. He saw that plants and animals living in climatically different parts of South America [from dry pampas to rain forest] were "incomparably more closely related to each other, than . . . to the productions of Australia or Africa under nearly the same climate" (*Origin,* page 347), as if the species adapted to different climates in South America had evolved from common ancestors on that continent. He saw also, when he reached the Galápagos, that, as he says (*Origin,* pages 397–398), on these islands "between 500 and 600 miles from the shores of South America [but much farther from any other continent] . . . almost every product of the land and water bears the unmistakable stamp of the American continent . . . ," as if the island plants and animals had evolved from American ancestors. And finally, what struck him most of all, he saw that slightly different species occurred on different islands of the Galápagos group. The surprise and importance of this discovery are expressed in his own words: (*Journal of Researches,* 1907 reprint of the 2nd ed., page 398):

I never dreamed that islands, about 50 or 60 miles apart, and most of them in sight of each other, formed of precisely the same rocks, placed under a quite similar climate, rising to nearly equal height, would be differently tenanted; but . . . I obtained sufficient materials to establish this most remarkable fact in the distribution of organic beings.

This was the checkerboard on which Darwin saw the game that evolution was playing.

Warblers, Flycatchers, and Ipswich Sparrow.   Persons who wish to do so can see the same kind of biogeographic evidence that Darwin saw, without going to South America and the Galápagos. Bird watchers can see a variety of wood warblers (family Mniotiltidae) and American flycatchers (Tyrannidae) in a variety of habitats in North America, but will find these birds absent in England and absent elsewhere in the Old World (except that two wood warblers have extended their breeding ranges from North America to the adjacent corner of Siberia; but they return to America to winter), the same habitats being occupied by warblers and flycatchers belonging to a different, chiefly Old-World family (Sylviidae). The explanation is, of course, that these different groups of birds have evolved from different ancestors in the New and Old Worlds respectively. Careful observers may recognize similarities of appearance and behavior among many wood warblers and American flycatchers, in contrast to the appearance and behavior of many sylviids. Or it may be necessary to go to books (for example, to Thomson, 1964) to confirm that these American and Old-World birds belong to different families. But having to go to books does not invalidate see-for-yourself biogeography. Darwin had to go to books for classifications and for much other information. Bird watchers can see too, for example, that the Ipswich Sparrow, which lives, or breeds, only on Sable Island more than 100 miles (180 km) off the coast of Nova Scotia, is very similar to the Savannah Sparrow of the adjacent mainland. Why should this be so, unless the island bird is derived from an adjacent-mainland ancestor?

Snail-Hunting Carabid Beetles.   Bird watchers in North America cannot easily see the pattern of differentiation of localized species that Darwin saw in the Galápagos, but entomologists can. For example, even amateur collectors can learn to recognize snail-hunting carabid beetles (tribe Cychrini, Figures 1 and 2), and can then discover by their own collecting that similar but slightly different species of one subgroup (subgenus *Steniridia*) occur on different mountain ranges in the southern Appalachians, slightly different species of another subgroup (subgenus *Scaphinotus* in a strict sense), on different mountains from southern

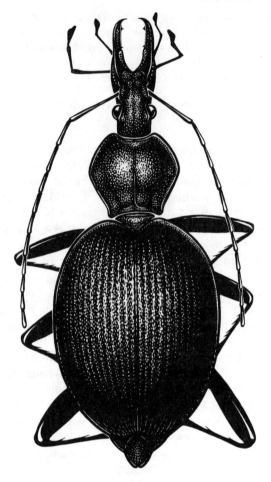

Figure 1. A snail-hunting carabid, *Scaphinotus (Stenoridia) ridingsi monongahelae* Leng, from Uniontown, Pennsylvania. Actual length, 17 mm.

Colorado to northern Mexico, and slightly different species of still another subgroup (subgenus *Brennus*), in different places from California to British Columbia. These beetles are flightless and almost as slow to disperse as the snails on which they feed. The only reasonable explanation of their present distribution is that the three groups of species have evolved from three different ancestors (which had evolved from an earlier common ancestor, probably a *Scaphinotus*), and that local speciation in each case has followed separation and isolation of local

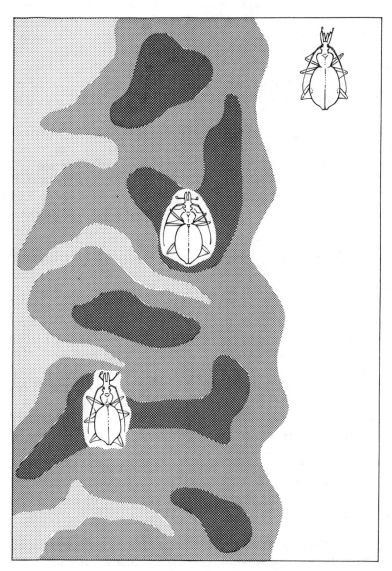

Figure 2. Occurrence of three species of *Scaphinotus* subgenus *Steniridia* in the southern Appalachians: upper right, *andrewsi* Harris, widely distributed at various altitudes; center, *aeneicollis* Beutonmuller, localized high on the Black Mountains; and lower left, *tricarinata* Casey, localized high on the Smoky Mountains. These species differ slightly in form and sculpture (in ways that cannot be shown in the small figures), but the decisive differences are in the male genitalia. See Valentine (1935) for further details and the distributions of additional species.

populations. In these cases, or at least in the first two, the separate mountains on which slightly different species occur are in effect islands for flightless insects adapted to the cool, damp mountain environment, and patterns of evolution are the same as on island archipelagoes. These beetles can still be collected, but some are rare as well as local. Most persons would do well to take the facts from papers cited at the end of this chapter rather than to risk overcollecting the small surviving populations.

This pattern—occurrence of basically similar but slightly different species on the different islands of an archipelago—really was decisive in leading both Darwin and Wallace to the idea of evolution. Darwin saw the pattern in the Galápagos; Wallace, in the Malay Archipelago. These men produced the "theory" of evolution, but they were led to it by geographic patterns that they saw in the real world, and that can be seen now.

The Pleistocene Eraser.   These geographic patterns do not now exist in England or parts of northern North America where ice sheets covered the land not much more than 10,000 years ago. The ice erased them, and there has not been time for evolution to make new ones. To emphasize this fact more forcefully, by analogy: the Galápagos Archipelago is like a chess board. Evolution has placed pieces (evolving plants and animals) on it in a definite pattern so that an intelligent man (Darwin) can see what game is being played. But in England and other northern regions ice sheets and climatic changes have swept the pieces off the board and thrown them back again almost at random, so far as the original "game" is concerned. The old evolution pattern has in fact been replaced by a new pattern of post-Pleistocene movement in the north. Darwin and Wallace had to get out of the shadow of the ice age and go to the tropics to see the game being played by evolution. This is more than an historical fact. Some characteristics and effects of evolution can still be seen only in the tropics, or by analyzing evolutionary patterns that involve the whole world (Chapter 7).

Summary of Geographic Patterns.   The geographic evidences of evolution—the patterns of geographic distribution that impressed Darwin in South America and the Galápagos and Wallace in the Malay Archipelago, and that can be seen now by naturalists who take the trouble—are usually ignored by critics who do not wish to accept the idea of evolution. This may be partly because most of the critics live in northern regions where Pleistocene ice and climatic changes have destroyed the most obvious evolution patterns. It is true that plant and animal distributions include striking discontinuities and anomalies, but these are reasonably explained by extinctions, which Darwin stressed. The fossil record shows a long succession of extinctions and replacements of dominant groups of plants and animals, and it would be surprising if some groups were not "retreating" now, and if their retreats did not sometimes result in discontinuities in the distribution of survivors. In any case, the irregularities do not change the fact that orderly geographic patterns made by evolution are the rule in most parts of the world. I have seen them in my collecting of carabid beetles, and some other animals, on islands of the West Indies and in Australia and New Guinea, as well as in unglaciated parts of North America. These patterns are a fact of life to naturalists, taxonomists, and biogeographers as well as to evolutionists, and there is no reasonable explanation of them except that they are being produced by evolution everywhere and at all times, although they are sometimes temporarily erased by cataclysms like Pleistocene glaciation.

Doubts and Criticisms

The fact of evolution is now accepted by most biologists and by many other persons who have had courses in biology or have read such books as *Evolution* in the *Life Nature Library*. However, the reality of evolution is widely disputed too, as the Scope's trial in 1925, the recent referendum on teaching of evolution in California, and many other incidents show. How evolution has occurred is disputed even more widely, by nonbiologists and even some biologists who do not like the reductionist frame (Chapter 10) or the idea of natural selection. Some criticisms of Darwinian evolution are reviewed in a readable book by Macbeth (1971). This is a forceful statement of objections to the current "theory" of evolution and criticisms of the evidence for it, but readers should be warned that Macbeth sometimes criticizes what he thinks evolutionists believe rather than what they do believe, and that his objections are partly political (as he himself says) rather than scientific. A more sophisticated attack on "Darwinism" has been made by the distinguished French zoologist Pierre Grassé (1973; 1977), whose criticisms are rebutted by Dobzhansky (1975).

Some actual and possible criticisms of the evidence of the fossil record and of geographic patterns are noted and answered in preceding pages. Logical or intellectual criticisms are more widespread and worth careful consideration.

Survival of the Fittest.   Macbeth repeats the old criticism of the concept of "survival of the fittest": that any organism that survives is fit by

definition, and that to say it survives because it is fit is circular reasoning. But evolutionists do not define "fit" this way. I do not think Darwin
used the word in the first edition of the *Origin*. What he did say was
(page 81) " ... can we doubt ... that individuals having any advantage,
however slight, over others, would have the best *chance* [my italics] of
surviving and of procreating their kind?" Modern mathematical evolutionists define and measure fitness *(W)* unrealistically (my Chapter 6A),
but they too are concerned with actual characteristics—details of genetic
composition, structure, function, and behavior—that determine relative
chances of survival. "Survival of the fittest" is a workable evolutionary
concept only because "fittest" is applied to individuals which, because
of their real characteristics, have the best *chances* of surviving in their
particular environments. Macbeth (page 62) says also that "survival of
the fittest ... is politically unacceptable. It smells of Hitler ... " This
statement is scientifically unacceptable. It smells of the Inquisition.

Probable Improbabilities.   Another criticism repeated by Macbeth, and
even by some biologists, is that it is *too improbable* that a process of
"random" variation and selection would produce, for example, man.
However, this is one of a class of situations in which the probability of a
particular outcome approaches zero, but in which the number of possible outcomes approaches infinity, and in which the probability that
some outcome will occur approaches one, or, in practice, certainty. For
example, if we could mark for identification enough water molecules
moving at random in the ocean to fill a thimble, and then dip up a
thimbleful of ocean water, the chance that we would get a set of molecules consisting of all the previously marked ones and no others would
indeed approach zero. But the number of possible combinations (sets)
of water molecules in the ocean does approach infinity, and the chance
that we would get some set of molecules which would fill the thimble
is, for practical purposes, certainty. This kind of probability can be expressed by an equation:

$$\rightarrow 0 \times \rightarrow \infty = 1$$

This might be called the probable-improbability equation. It does not
mean that a probability that approaches zero in single cases, multiplied
by a number of cases that approaches infinity, must give a probability
that approaches one, but only that it does in fact do so in many actual
situations.

   This kind of probability occurs in some biological processes. For example, the genes that make up a human individual are so numerous,

and they are so effectively shuffled in bisexual reproduction, that it is very improbable that any two individual humans have ever had or ever will have identical sets of genes (except identical twins, who share the same gene sets). It is exceedingly improbable that *any* particular set of genes will occur. But the number of possible sets is exceedingly great, and each normal human being does receive a set. (Abnormal gene sets are not the result of failure of the probability formula, but are caused by other factors.)

To apply this concept to, for example, man: it was probably inevitable that an intelligent organism in some ways like man would evolve on earth. The probability that this organism would be man as he is, in all details, approaches zero. But the possibilities approach infinity, and one of these possibilities was (practically) sure to occur, and man is in fact the one. That evolution has produced man is not more unlikely, or at least involves the same kind of probability, as that, when we dip up a thimbleful of sea water, we will get a set of water molecules sufficient to fill the thimble.

Monkey and Typewriter. The kind of probability involved in organic evolution can be described by a revision of the well-known fiction of the monkey and the typewriter. (The history of this notion and some modifications of it are suggested by Bennett, 1977.) This usually takes the form of a simple statement: a monkey striking the keys of a typewriter at random will, if he keeps on long enough, write all the books in the British Museum, by the laws of probability. Now this *is* an unrealistic fiction. Suppose we consider only the 26 letters of the alphabet, ignoring punctuation and spacing, and only one small book of, say, 20,000 words averaging 5 letters each. And suppose that, whenever the monkey strikes a wrong letter, he must start again at the beginning of the book. The probability then will be only something like $26 \times 10^{-100,000}$ that the monkey will write this particular predesignated book with one start, and he will have to make something like $26 \times 10^{100,000}$ (26 with about 100,000 zeros after it) starts if he is to have a reasonable chance of writing the book at all. (These are rounded figures which are not mathematically precise.) The probability that the monkey will write even this one small book in any conceivable time is so small that the idea is, for practical purposes, a fiction. If evolution of life on earth depended on probabilities of this order, we would still be inorganic atoms.

Now, suppose that whenever the monkey strikes a wrong letter the letter is erased and the monkey allowed to go on striking at that point until he hits the right letter. Then he will write the predesignated book in one start and something like $26 \times 100,000$ strokes, and if the monkey is a reasonably fast striker, he will write the whole book in a few days. But he can do this only because wrong letters are eliminated by an external intelligence or by reference to a pre-existing model, and biologists do not detect this kind of intelligence or model in evolution.

But now, suppose that the monkey, still striking the typewriter keys at random, produces a sequence not of inert but of living letters—bits of living matter—and suppose the bits are selected (selectively erased) as a result of their own characteristics, behavior, and relation to other letters in the sequence. And suppose that selection erases not only nonviable single letters as they are struck but also in the course of time relatively inefficient groups and sequences of letters—words, sentences, paragraphs, and chapters. Then, provided the additions are greater than the erasures, the probability may approach one (certainty) that a coherent book will be written. It will not be a predesignated book; the general direction of it (i.e., the general subject) may be predicted, but not the details—this, of course, is true of real books written by human authors as well as of evolutionary "books"; and the time required may be predicted too, but only within wide limits.

This is the kind of situation that obtains in evolution: the source of energy (analogous to the monkey) is the molecular and chemical energy of atoms and molecules; the "living letters" are genes and their components, mutants, and combinations; and the additions have in fact been greater than the erasures, for otherwise evolution would not have occurred. Under these conditions, the probability that evolution will produce any pre-designated organism approaches zero, but the probability that some organism comparable to, say, man in organization, complexity, and intelligence will evolve may approach certainty. The time will depend on many factors, but the three billion years or so since life appeared on earth would seem sufficient for the evolution that has in fact occurred.

Summary

The clearest evidences of the fact of evolution, the evidences that first impressed Darwin and that anyone can see, are the geographic patterns made by evolving plants and animals on the earth's surface. The patterns can hardly be explained except as products of an ongoing evolution which gradually produces change and diversification among all living things everywhere. The fossil record is decisive too, but not so visible to most observers. And additional evidence comes from virtually every subfield of biology; modifications of domestic animals and plants under selection by man, and detailed homologies revealed by comparative anatomy are perhaps most easily appreciated. Full acceptance of the fact of evolution requires a satisfactory explanation of the mechanism; most of the rest of this book is concerned with this. Gaps and anomalies occur in the factual record, but are to be expected. Logical and intellectual objections to evolution are based at least partly on misunderstandings, including misunderstanding of what biologists mean by "survival of the fittest" and of the kind of probability involved in evolution.

Webster's Collegiate defines proof as "that degree of cogency, arising from evidence, which convinces the mind of any truth or fact and produces belief . . . " Different minds will require different "degrees of cogency," but I think that most persons who look at the evidence for themselves, and are not prevented by religious or political prejudices (i.e., by judgments before the evidence) will accept evolution as a fact. And I think that most readers will find the explanation of it satisfactory, with the provisos that many details and probably also some important generalizations are still to be discovered and that, although the principles are simple, their manifestations in the real world are often inconceivably complex.

READING AND REFERENCES

Darwin, C. 1839. *Journal of Researches*. . . . (See my Chapter 1.)

Darwin, C. 1859. *The Origin*. . . . *(See* my Chapter 1.)

Gould, S. J. 1977. *Ever Since Darwin*. New York, Norton. (A set of essays of the sort that evolutionists ought to write, although some are marred by what seem to me to be political prejudices.)

Variation Under Domestication

Mangelsdorf, P. C. 1974. *Corn. Its Origin, Evolution, and Improvement*. Cambridge, Massachusetts, Harvard Univ. Press.

Comparative Anatomy

Forelimbs

de Beer, Sir G. 1971. *Homology, An Unsolved Problem*. Oxford Biology Readers, No. 11. Oxford, England, Oxford Univ. Press; now available from Carolina Biological Supply Co., Burlington, North Carolina 27215.

The Human Hand and Foot

Napier, J. R. 1976. *The Human Hand.* Oxford Biology Reader, No. 61. Oxford, England, Oxford Univ. Press; now available from Carolina Biological Supply Co., Burlington, North Carolina 27215.

Reynolds, V. 1967. *The Apes.* New York, Dutton.

The Fossil Record

Romer, A. S. 1959. Darwin and the Fossil Record. *Natural History,* **68**, 457–469.

Simpson, G. G. 1951. *Horses.* Oxford, England, Oxford Univ. Press; and 1961, New York, Am. Museum Nat. Hist.

Various Authors. 1978. *Evolution and the Fossil Record. Readings from Scientific American.* San Francisco, California, Freeman.

See also Chapter 8.

Geographic Patterns

Darlington, P. J., Jr. 1959. Darwin and Zoogeography. *Proc. Am. Philos. Soc.,* **103**, 307–319.

Warblers, Flycatchers, and Ipswich Sparrow

Thomson, L. 1964. *A New Dictionary of Birds.* London, England, Thomas Nelson, for British Ornithologists Union.

Snail-Hunting Carabid Beetles

Valentine, J. M. 1935. Speciation in *Steniridia,* a Group of Cychrine Beetles. *J. Elisha Mitchell Sci. Soc.,* **51**, 341–375, Pls. 65–73.

Ball, G. E. 1966. The Taxonomy of the Subgenus *Scaphinotus* Dejean . . . . *Trans. Am. Entomol. Soc.,* **92,** 687–722, Pls. 54–58.

Gidaspow, T. 1968. A Revision of the Ground Beetles Belonging to *Scaphinotus,* Subgenus *Brennus. Bull. Am. Museum Nat. Hist.,* **140**, 131–192.

Doubts and Criticisms

Macbeth, N. 1971. *Darwin Retried.* Boston, Massachusetts, Gambit.

Grassé, P-P. 1977. *Evolution of Living Organisms. Evidence for a New Theory of Transformation.* New York, Academic. (Transl. of 1973 French ed.)

Dobzhansky, T. 1975. Darwinian or "Oriented" Evolution? *Evolution,* **29**, 376–378. (Review of the French ed. of Grassé.)

Monkey and Typewriter

Bennett, W. R., Jr. 1977. How Artificial Is Intelligence? *American Scientist,* **65**, 694–702.

3

METHODS: LIMITS AND COMPROMISES; MODELS, ANALOGIES, AND MATHEMATICS; SIMPLIFICATION; SEMANTICS; CARABID BEETLES

This chapter is a personal statement of the methods I shall use to try to describe and elucidate organic evolution.

Evolution cannot be observed directly, except perhaps small fragments of it under special circumstances. Evolutionists have to grope more or less blindly toward understanding of it, and different evolutionists see it in different ways: as descent with modification in sequences of individuals, as statistical changes in gene pools and populations, or as nucleotide substitutions in DNA; or as a stochastic process, as adaptation to the environment, as change programmed at least partly by internal forces, or as a response to mystical perfecting forces. There is partial truth in all these concepts, except I think the last, but none gives a full and balanced view of evolution. The biologist with the best perspective on evolution as a whole is, I think, the naturalist who looks directly at what is going on in the woods, fields, and waters around him, provided he keeps up reasonably well with at least the general trend of new discoveries and theoretical advances at all levels from molecular genetics to the theory of faunal equilibriums. My method is primarily the naturalist's: whenever I can I shall derive and test hypotheses and models of evolution by what I see, or think I see, in the real world, and I shall try to describe what I see in words that will make evolution a visible reality to naturalists.

Limits and Background

We must, I think, try to describe and explain evolution within practical limits. We must accept: matter as we find it on the earth's surface, regardless of theories of its origin; the conservation of matter and energy, expressed in the first and second laws of thermodynamics; and the direction of time. We must accept Newtonian physics and, for the time being, forget relativity and the "new physics" including the "uncertainty principle," which edges into indeterminism if not into mysticism. And we must accept "fact" and "reality" in their everyday, not metaphysical, meanings. In short, we must at least try to see evolution as an observable, explicable process, occurring in a real world which is what it seems to us to be. From this point of view, man is an animal produced by evolution, and we must look for evolutionary explanations of all man's characteristics, although the explanations need not be simple.

This is to say that, if we wish to understand evolution, we should begin by adopting a limited, practical, mechanistic reductionism. We should accept the world we can see and measure, and make the most of it. Evolutionists who wish to go beyond this, into vitalism, mysticism, or metaphysics, may do so, but they should make very clear when they change from strictly scientific to less restrained modes of thought. This point is elaborated in Chapter 10.

Compromises

Several compromises have to be made in the writing of any book on evolution.

Coverage.   There must be a compromise on coverage: to cover the subject as a whole, parts must be treated briefly, by concise summaries, with references to more-detailed treatments. In the present book, I shall begin (because I must) by assuming that readers have, if not some initial knowledge of biology, at least a good, modern, biological textbook for reference, which includes something about classification of plants and animals and the characteristics of the major groups. Then, I shall give summary treatment to aspects of evolution that have been well covered recently, including the origin of life, molecular genetics, and the mathematical theory of selection, but shall treat in more detail aspects of the subject that are usually not so well covered, including the cost of evolution and the imprecision of adaptation, multi-level selection including "group selection," and geographic aspects of evolution. Even in these cases, however, the amount of detail that can be given is limited, and treatments of both situations and processes will have to be simplified. (The kinds, uses, and misuses of simplification are discussed later in this chapter.)

Histories of Ideas. Another compromise has to be made in tracing histories of ideas. Some modern evolutionists seem not to be interested in the history of ideas at all, but propose as if new ideas that date from Darwin. I think that present theory should at least be put into historical perspective. But limitation of space prohibits the tracing of ideas in detail. My compromise is to stress Darwin's ideas, which were extraordinarily comprehensive as well as historically decisive, and through Darwin to emphasize the continuity of essential evolutionary concepts from the premathematical, premolecular era of biology to the present, but otherwise to summarize the histories of evolutionary ideas very briefly. All evolutionists should profit from a careful reading or rereading of the *Origin,* in which Darwin states nonnumerically but clearly and often forcefully some very modern ideas about, for example, competition, extinction, faunal equilibriums, and the relation of evolution to geography, including the apparent relation between area and faunal dominance.

Literature. A double compromise has to be made in coverage of the literature on evolution, first in the reading that one man can do, and second in selecting references to cite. This is a painful compromise for me. In my taxonomic work on beetles I am required to find and cite every pertinent publication since 1758. Writers on evolution have no such convention, and cannot have; too much has been published. What I do is sample current publications in much the way that a population biologist samples a population, attempting to obtain a sample that represents the whole reasonably accurately. I read the book lists and book reviews in *Science, Nature, The Quarterly Review of Biology, The American Scientist, Scientific American,* and *Science Books;* buy books that seem worth having; take and read several journals, including *Evolution;* refer to other journals on occasion; and ask biologists about recent publications in their special fields. (I rarely use abstracting journals; I found them a waste of time in taxonomy and think my time is better spent sampling originals than collecting abstracts; but I wish more authors would include abstracts with their originals!) From the several thousand items accumulated in this way during the last ten years, I have read a few carefully, have skimmed or sampled many more, and have shelved or filed the rest, entering them in a subject index to help me find them again.

Reading and Reference Lists. This book cannot be documented like a research paper. The chapter reference lists are intended only as stepping-stones to further reading, and to validate quotations and critical statements in the text. The references are arranged by subjects under a sequence of heads corresponding to that in the text, but the sequence is abridged by omission of heads under which no references are given. My intention (not always fulfilled) has been in each case to cite both a good general work that naturalists should get pleasure as well as profit from reading, and at least one recent book or paper to bring the subject up to date. The references are not always chronological but are in some cases arranged according to their importance or their sequence in the text, but the works cited under one heading are so few that readers should have no trouble finding them. The informality of the lists is deliberate, the result of thought and experiment; lists like these would not be suitable for a more formal text, but do fit this book.

Models

All that we know of reality is in the form of models. This is a practical, not metaphysical, fact. What our brains cannot model we cannot conceive of or understand; for example, because our brains are finite and can make only finite models, we cannot conceive of infinite space or infinite time. We can represent these things mathematically, but this is not the same as conceiving of them intellectually. Our inability to conceive of infinity is not the point now, however. The point is that, in making models of evolutionary processes, we are doing only what the human brain must do in any case.

The general model of multi-level evolution that will be used here consists of sets. A set at any level consists of two or more units, each itself a set of smaller units, that occur together or within interacting distances, that do or may interact, and that do or may undergo directional change. Models like this are considered in more detail in Chapter 4. They are *particulate* models which fit well with reality. Other kinds of models are derived from analogies and mathematics.

Analogies. An *analogy* is a "relation of likeness" between things that are alike not in themselves but in their attributes or actions. Models of evolution are analogies, and this is true whether the models are verbal or mathematical. Verbal analogies, or analogy models, are very useful in describing evolution. Darwin used them freely: his comparison of the evolution and dispersal of organisms on the earth's surface to the growth of a tree (quoted at the beginning of my Part Three) is one, and another is his description of the large northern continents as "efficient [evolutionary] workshops of the north" (my Chapter 7). Molecular biologists use them: the comparison of DNA to coded information is an analogy (see end of my Chapter 5). I shall use verbal analogies freely.

Evolutionary analogies are of several grades. The concept of evolution as directional change in sets is hardly an analogy at all but a description of the actual process. True analogies may be either close or distant. (These are my terms, not in general use.) Close analogies involve things that are related in themselves as well as in their attributes or actions. For example, comparisons of an insect or human society to a multicellular individual, and of both to an evolving population of nonsocial individuals, are close analogies, for societies, individuals, and populations have all evolved from similar ancestral sets by set selection (Chapters 5 and 7). Distant analogies involve comparisons of things that are not related in themselves but are more or less similar in their attributes or actions. My comparison of the force of selection to the force of an hydraulic ram (Chapter 6A) is a distant analogy, and so is the comparison of a set of genes to a printed message. And mathematical models of evolution are distant analogies too.

One more point about analogies, whether verbal or mathematical: they are aids to understanding, but no matter how pertinent and striking they may be, *they are not evidence;* they tell us what to look for, but not what we will necessarily find when we look.

Mathematics.   Most models of evolution now are mathematical, and the mathematical models are too-often accepted with an uncritical respect they do not always deserve. I hope, therefore, that readers will consider seriously the following essay on use of mathematics in the study of evolution. It is a careful assessment, not a hasty tirade.

Before coming to actual use of mathematics, some propaganda needs to be cleared away. Even good biologists sometimes say (quoting physicists) that "unless you have measured it [and can apply mathematics], you don't know what you are talking about." This is propaganda. Do persons who say this mean that no nonmathematical biologists from, say, Darwin to W. M. Wheeler knew what they were talking about, and that as soon as measurements were made and mathematics applied biologists did know what they were talking about? Few thoughtful persons would agree with this! The quotation *is* propaganda, and it can only be answered by counterpropaganda. For example, we can say that, unless you know what you are measuring, you don't know what you are talking about, and this is true as well as fair counterpropaganda. (If you must measure it before you can know what you are talking about, but cannot measure it until you know what you are talking about, you can never know what you are talking about, but this is metaphysics and not biology.) And we can, if we wish, carry counterpropaganda further and recall Trollope's (in *The Three Clerks*) "Mr. Minusex, ... who never thought but always counted"; fortunately few mathematical evolutionists are as bad as this. And we can quote Bundy (1970, page 351) on the tendency of modern theoreticians (not just evolutionists) to "slip off into a kind of scientism which brings an appearance of precision by the abandonment of reality," and this is a real and dangerous modern trend, not just a counterpropaganda bugbear.

To turn from propaganda to reality, there is a conspicuous dichotomy in the reality and predicting power of mathematical models. We see this in what goes on around us from day to day. The precision with which mathematics and computers can put men on the moon, within a few 100 m of a predetermined point at a range of something like 400,000 km or $\frac{1}{4}$ million miles, is an example of the reality of mathematical models when all significant factors are known and can be precisely quantified. On the other hand, failure of mathematics and computers to predict weather accurately even at short range is an example of the unreality of mathematical models when significant factors are not all known or cannot be precisely quantified.

This dichotomy extends to mathematical models of evolution. The precision with which they can describe and predict simple Mendelian processes or the outcomes of population-growth and competition experiments in glass jars proves the reality of the models in artificially simple cases in which factors are few, known, and quantifiable. On the other hand, the factors concerned in real cases of organic evolution are more numerous and more difficult to identify and measure than the factors that determine weather. Mathematical models of real evolutionary processes may therefore be even more wrong than weather predictions. But while bad weather predictions are promptly falsified by visible events, bad mathematical models of evolution are not, because evolution cannot be seen as weather can. So, erroneous models of evolution may be uncritically accepted. This is not just a remote possibility but a frightening probability, frightening because the wrong conclusions are likely to be applied to man with consequences incomparably more disastrous than having a picnic rained out.

The suggestion that mathematical models may not fit reality is too important to dismiss unsupported. Examples are required to clarify and emphasize it. The first example, visible especially to naturalists, is taken from a paper by two mathematical ecologist-evolutionists who have deservedly high reputations. May and MacArthur (1972) present a mathematical model which they say shows that "there is an effective limit to niche overlap in the real world," by which they mean that species competing for a "one-dimensional resource spectrum" *must* divide it into not-too-widely-overlapping fractions. But naturalists have only to look out the window on a summer day to see hundreds of species of green plants coexisting in life-and-death competition for sunlight (literally a "one-dimensional resource spectrum") without dividing it as the model requires. Surely this should shake the confidence of any thoughtful naturalist in the predictions of even the most highly regarded mathematical evolutionists! The model may hold in special cases, but it does not justify May and MacArthur's sweeping generalization. And even in special cases, resource division may be a "convenience" rather than a necessity, as among the apple pickers described in Chapter 6A under *"Competitive extinction and exclusion."*

Simpler cases emphasize how unreal mathematical predictions can be. A keen sense of smell and instinctive ability to swim should have been advantageous to our ancestors for millions of generations. Numbers of individuals in every generation must have been selectively eliminated because they failed to smell hidden food or enemies, or could not swim; some humans are still eliminated prematurely because they do not smell smoke or intruders in time, or cannot swim. Simple arithmetic predicts that selection should have "maximized" a keen sense of smell and instinctive swimming ability in man, but it has not; possible reasons why not are suggested in Chapter 6B, under *"Cost and the imprecision of adaptation."* And kin selection too is based on arithmetic so simple and forceful that most evolutionists accept it uncritically. The theory, or part of it, is that if one brother saves another from drowning and enables him to reproduce, the one, in effect, hands on half his own genes to the next generation, and brothers should therefore be under strong, genetically determined compulsion to save each other. But if one brother lets another drown and reproduces in his place, the one hands all his own genes instead of only half, and brothers should be under even stronger compulsion to let each other drown. How can mathematicians ignore this second, simple calculation but accept the more complex and less profitable calculations of kin-selection theory, as they do? Kin selection is criticized in detail in Chapter 7, with the conclusion that it is simplistic and unrealistic, except in very special cases. I suspect that it has "mousetrapped" more good evolutionists—led them into untenable positions—than any other current evolutionary idea.

Evolutionary mathematics seems to rest on a usually unstated three-step logic: (1) Adaptations evolve by selecton. This is true. (2) Mathematical models can predict the outcomes of simple, separate selective processes. This is true too. (3) Predictions that are good in simple, separate cases are good also in complex real situations. But this does *not* follow. The predictions may be spoiled by genetic programming (Chapter 6B); or by the cost of selection (Chapter 6B); or (in the case of brain-determined behaviors) by competition for space on the brain (Chapter 6C, under *"Specialization and focusing");* or by competition among adaptations, which may inhibit each other (as, for example, more-profitable population-wide-reciprocal altruism may inhibit kin-limited altruism; Chapter 7). These and perhaps other unknown factors interact, change as complex environments change, and all together probably often do greatly change, inhibit, or reverse selective processes in complex real situations, making mathematical predictions about evolution in fact as unreal as weather predictions often are.

Of the enormous amount of evolutionary mathematics that is published now, much is repetitive and much is in fact unrealistic. In saying this I do not mean to deny the value of mathematics, but only to protest its misuse and urge its proper use. Many evolutionists of the older generation, to which I belong, have surely undervalued mathematics; we are committed to nonmathematical methods. Biologists of the present middle generation, now actively publishing and teaching, often overvalue it; they are committed too, and are probably beyond the reach of criticism. But I hope the next generation, perhaps now at the graduate student level, will put mathematics in its place; will value it and use it; but will insist that mathematical evolutionists define their terms precisely, describe their models completely, and relate them to the world realistically. There are risks in doing this. It is easier and safer to make mathematical models than to attempt to study evolution in the real world: easier to make an equation or program a computer than to measure factors affecting evolution in nature, and safer because (present fashion being what it is) a mathematical treatment almost guarantees a result that can be published whether it is right or wrong, while an attempt to analyze a real situation may end in frustration and failure to reach any publishable conclusion at all. And some professors of biology may not tolerate students who are critical of the professors' math, so that innovative students may risk their degrees or appointments.

Since I have hoped that the next generation of biologists will put mathematics in its place, I had better say what I think its place is. Scientific work consists of theory and observation, which are complementary: theories are formulated, then tested (falsified or not) by observations, then further theories are formulated, and tested by further observations, and so on. This is the "hypothetico-deductive" process, reviewed in a readable way by Medawar (1969). Observations consist of observations of natural phenomena supplemented by observations of material models (experiments). And theory consists of intellectual, verbal hypotheses, supplemented by mathematical models. Most evolutionary mathematics is theory, based not on observation of evolution but on theoretical postulates which are often incomplete and sometimes erroneous. And the mathematical treatments add possible mathematical errors to any omissions or errors that occur in the postulates. The place of mathematics, then, is that of second-stage theory, which should come after intellectual theory and is dependent on the latter's completeness and accuracy, and which is therefore two steps removed from reality. It should be tested against reality, never mistaken for it. Putting it in its place is part of the return to reality that I think will characterize the next stage in the evolution of evolution theory (Chapter 1).

In practice, cases should always be formulated conceptually and verbally before they are quantified and treated mathematically. This is not only logically the right way, but often the way that results in innovations. For example, several botanists before Mendel had noticed dominance and segregation of characters in successive generations of some plants (Dunn, 1965, pages 25–32), but Mendel was the first to whom it occurred—*before* he began to count his peas—that segregation might be quantitative, and it was this preliminary idea that led him to the discovery of what we now call Mendel's laws. And, after quantification and mathematical analysis, conclusions and their significance should be stated in words, simply but fully, if the conclusions are to be widely and usefully understood. Mendel did not do this. He did not extend his findings to a generalization about the role of Mendelian heredity in evolution. If he had, and had stated the generalization clearly, his discovery would not have had to wait a third of a century for recognition. (In fact, Mendel did not understand the evolutionary significance of his findings.)

Another general comment on evolutionary mathematics: if adaptations evolving by selection in natural populations are so numerous, complex, costly, and slow, and if environments are so complex and change so rapidly, that most organisms most of the time are far from perfectly adapted (Chapter 6B), the assumption routinely made by mathematical evolutionists that selection must have "maximized" such things as reproductive strategies, foraging patterns, and (especially!) social behaviors is wrong. And if the assumption is wrong, a large part of recent, "sophisticated," evolutionary mathematics is wrong—widely divorced from reality.

In summary, most mathematical models of evolution are concerned only with parts—single parts of evolving organisms and single factors in environments—and when the parts are put together into complex wholes, as they are in the real real world, the mathematical models often fail, sometimes dangerously.

Simplification; simplistic

The following is a consideration of methods of simplification and of their uses and misuses.

Two kinds of simplification can be distinguished. *Simplification-by-synthesis* results in simple statements—generalizations—which cover all cases and are true of all cases. An example is the statement that selection is differential elimination of pre-formed sets; it is true of all cases at all levels. *Simplification-by-omission* results in statements which are not true of all cases. An example is the statement that selection is differential reproduction; it is true only at the level of living organisms, and is only part of the truth at this level.

Both kinds of simplification have their uses and misuses. *Simplification-by-synthesis* is a useful and sometimes essential means of making complex reality comprehensible. It is misused only when the simplifications are not true or when actual complexities are not acknowledged. *Simplification-by-omission* is useful to reduce complex situations and processes to forms that allow preliminary analysis. It too is misused when actual complexities are not acknowledged.

A pertinent proposition is that discovery that a thing is more complex than it seemed at first does not make it a new thing. If we see a rain forest from the air, it seems a coherent whole. If we then enter it, we discover that it is in fact complex. But it is still the same coherent whole, and for some purposes is best thought of as such. Similarly, Darwin saw evolution as a process of variation and selection among individual organisms in populations. Now we have discovered that the process is complexly modified in many ways. But it is still essentially the same process of variation and selection, and for some purposes is best thought of simply as such, although its modifying complexities should not be forgotten. The current theory of evolution is sometimes called "synthetic" or "neo-Darwinian," but I dislike these terms because they overemphasize complexities and hide the historical continuity of what I prefer to call the Darwinian theory.

Another proposition is that fundamental principles, derived by simplifications-by-synthesis, are usually not changed by discovery of new details no matter how exciting the latter may be, while models derived by simplification-by-omission are frequently changed or falsified by new discoveries.

Some biologists now seem almost automatically to condemn all simplification as "simplistic"; they use this word almost as loosely as some politicians use "communist." But *simplistic* is defined (in *Webster's New Collegiate Dictionary*) as applying to "the act or an instance of oversimplifying; esp: the reduction of a problem to a false simplicity by ignoring complicating factors." This definition does not fit simplifications-by-synthesis which are true and useful and which are acknowledged to be simplifications of more-complex reality. It does fit simplifications-by-omission which are presented as if they represented reality. Much evolutionary mathematics is literally simplistic by this definition.

The present book is, in part, an essay in simplification-by-synthesis. It is an attempt to reduce evolution to simple principles and to put the principles together into a coherent whole, without hiding but continually stressing the complexity of reality. I shall try especially to remind readers that explanations may be complex. The principle of parsimony, or of Occam's razor, which is that simple explanations are preferable because they are simple, should be used only to make exploratory cuts into problems, not to exclude complex explanations.

Speculation

Any fool can speculate, as Darwin said, but speculation is not therefore foolish. Darwin was himself a prodigious speculator. Some scientists condemn speculation and seem to think that science proceeds entirely by observation and inference from observation, but this is not the way the human mind works. Ideas, including theories of evolution, evolve by process of innovative speculation and selective elimination—the hypothetico-deductive process, already referred to. We imply the importance of speculation when, reading someone else's paper, we say, "I wish I had thought of that!" Scientists *should* speculate, as I shall do. But scientists should clearly distinguish speculation from observed fact, and also should monitor their own speculations and sort out unrealistic ones, as Darwin did, and as I shall try to do.

Semantics

Semantics is "the science of meanings," or of the transfer of information especially by words.

Evolutionists presumably want to transfer information not only to each other but also to interested laymen. They should want this. Understanding of organic evolution ought to be made as easy as possible for everyone who needs it for use in or out of science, in studies of social processes and of the nature and problems of man. Evolutionists should therefore plan their words to facilitate broad understanding. This idea is familiar to the point of triteness, and (to be trite myself) familiarity seems to have bred contempt; the idea is most often honored in the breach.

Of course this is true among biologists in general, not just among evolutionists, who must use the same words, defined in the same ways, that other biologists use. And of course special words are necessary in biology and evolution. But they should be kept as few and as easy to understand as consistent with precision. Definitions should be simple and precise. And, most important of all, everyday words should be used as nearly as possible in their everyday meanings, and when they are used in special ways, they should be defined with special care *at the time they are used.* Everyday words used in special ways by biologists and in different ways by different biologists or by the same biologists at different times include "dominance," "competition," "selection," "niche," "intelligence," "race," "species," and "information." These words and others like them will be defined here, not in a glossary (which is not needed), but at appropriate places in the text, referred to in the index.

Some semantic implications are so built into our everyday language that they are difficult to avoid, notably implications of purpose or teleology and of anthropomorphism. For example, we may say "wings are to fly with," and this seems to imply purpose, but to most evolutionists it means only that the function of wings is flight and that flight gives wings the selective advantage that results in their evolution. This implies cause but not purpose. Most evolutionists see no purpose in evolution even when they use language that seems to imply it. I shall try not to use teleological language and (if I use it inadvertently) shall never intend to imply purpose by it.

Carabid beetles

I shall use carabid beetles (family Carabidae, "gound beetles,") (see Figure 1 in Chapter 2 and Figure 4 in Chapter 6A) to illustrate several evolutionary patterns, so they had better be introduced here. They are typically ordinary-looking, active, ground-living, predaceous beetles (but actually much more diverse than this especially in the tropics), characterized technically by division of the first ventral abdominal segment, legs adapted for running or digging (not swimming), wings (if fully formed) with characteristic venation, and many other details that need not be given here. About 30,000 species of them are known, more than of all vertebrates together, excepting marine fishes. Different species range in size from less than 1 to about 70 mm in length, and they differ also in ecologic requirements, special adaptations, and behaviors. Naturalists see them everywhere, and can easily learn to recognize them.

Three important geographic patterns related to evolution are well illustrated by carabids in North America. One is a pattern of geographic differentiation of species. It is the same kind of pattern that Darwin saw on the Galápagos, but the Galápagos are hard to reach and hard or impossible to collect on now, and some of the animals that impressed Darwin (some of the giant tortoises) are extinct, while flightless carabids that show similar local-differentiation patterns still exist in reasonably accessible parts of North America.

A second evolutionary pattern exhibited by carabids has to do with their wings and flight. They are all descended from winged, flying ancestors, and most existing species are still winged. However, the wings have atrophied and the insects therefore become flightless in many different stocks in many different places. Their wing atrophy apparently begins by mutation from a long- to a short-winged condition, and mutation often results in wing dimorphism, with long-winged flying individuals and short-winged flightless ones occurring together in the same populations. In the simplest cases, the alleles of a single gene determine whether or not individuals can fly. This invites strong selection, and the distribution of long and short-winged individuals shows clearly that selection has occurred, its direction depending on the situation in each particular case. Situations have in some places favored flight and in others flightlessness (Chapter 6A).

A third evolutionary pattern exhibited by carabids is one of formation of faunules (small sets of species) localized in special places. Because carabids are numerous in species almost everywhere, because they are diverse ecologically, and because they include flightless as well as flying species, the faunules that evolve on mountain tops and islands (and in caves, although these are not considered here) can be analyzed statistically, and their origins and evolutions can often be traced with confidence (Chapter 7).

These evolutionary patterns shown by carabids exist in nature, in the real world now, and can be seen and studied there by any reasonably competent naturalists who know a little about insects, even by amateurs. However, carabids are not so suitable for study in the laboratory. They breed slowly: most of our species have only one generation a year. And the larvae are often cannibalistic and have to be reared individually, one to a tube, not many together like *Drosophila*.

Summary

Evolution is described here by a naturalist writing for other naturalists. Reality is taken as what it seems to us to be, and life and evolution are seen from the point of view of a practical, mechanistic reductionism. The method is hypothetico-deductive; concepts and models are proposed, then tested against the real world, so far as possible; but this procedure is usually implicit rather than formal. The models are primarily intellectual and verbal rather than mathematical; analogies are used freely; and mathematical models are criticized and the hope expressed that in the future they will be used more realistically than now. An effort is made to use plain statements that are literally true, unambiguous, and as simple as consistent with reality, and to distinguish clearly fact, probability, and speculation. Style and vocabulary are calculated to be not only scientifically explicit but also intelligible to laymen as well as biologists. New examples of evolutionary patterns are taken from carabid beetles; the examples can be seen in accessible parts of North America by naturalists willing to take the trouble.

READING AND REFERENCES

Limits and Background

See Chapter 10, "The reductionist frame," and Chapter 4, "The cosmic theater."

Compromises

Coverage

Wilson, E. O., et al. 1973. *Life on Earth*. Stamford, Connecticut, Sinauer Associates. (The text I use.) (There is a second edition, 1977)

Various authors. 1976, 1977, 1978. [Reviews and comparisons of some other textbooks of biology.] *Quart. Rev. Biol.*, **51**, 96–101; **52**, 286–288; **53**, 425–426.

Histories of Ideas

See Chapter 1.

Models

Mathematics

Bundy, M. 1970. Were Those the days? *Daedalus*, Summer.

May, R. M., & R. H. MacArthur. 1972. Niche Overlap as a Function of Environmental Variability. *Proc. Nat. Acad. Sci. USA,* **69**, 1109–1113.

Darlington, P. J., Jr. 1972. Competition, Competitive Repulsion, and Coexistence. *Proc. Nat. Acad. Sci. USA,* **69**, 3151–3155. (Includes a criticism of May & MacArthur's paper.)

Medawar, P. B. 1969. *Induction and Intuition in Scientific Thought.* Philadelphia, Pennsylvania, Am. Philos. Soc.

Dunn. L. C. 1965. *A Short History of Genetics. (See* Chapter 1.)

Carabid Beetles

Ball, G.E., et al. (eds.). 1979. *Carabid Beetles: Their Evolution, Natural History, and Classification.* The Hague, W. Junk.

Ball, G. E., 1960. Carabidae. (In) Arnett, R. H., Jr., *The Beetles of the United States,* Ann Arbor, Michigan, Am. Entomol. Inst., pages 55–174. (The best—most natural and useful—classification of groups within the family Carabidae.)

Lindroth, C. H. 1961–1969. The Ground-Beetles (Carabidae, excluding Cicindelinae) of Canada and Alaska. Six parts, as supplements of *Opuscula Entomologica;* for details *see Systematic Zoology,* **20**, 367–368. (A realistic, usable taxonomic treatment of the species of the northern half of North America.)

Thiele, H-U. 1977. *Carabid Beetles in their Environments.* Berlin, Germany, New York, Springer-Verlag; reviewed by Cooper, *Quart. Rev. Biol.,* **54**, 93–94. (The first comprehensive, but still very incomplete, book on Carabidae as living animals.)

PART TWO

Processes and Levels of Directional Change

Thus the whole is complete on one principle. The masses of space are formed by law; law makes them in due time theatres of existence for plants and animals; sensation, disposition, intellect, are all in like manner developed and sustained in action by law. It is most interesting to observe into how small a field the whole of the mysteries of nature thus ultimately resolve themselves.

Robert Chambers, *Vestiges of the Natural History of Creation,* 1844 (quoted from the reprint of the first edition, pages 359–360).

4

THE ESSENTIAL PROCESS:
SETS AND SET SELECTION

This chapter is concerned with the theory of multi-level evolution, or with the essential characteristics of directional change at all levels from the simplest nonliving to the most complex living ones. It is an attempt to reduce complex reality to terms that are simple but not simplistic (see Chapter 3). It may be the most novel chapter of this book—but parts of it have been published previously in short papers "spun off" as the book was being written.

The theory of multi-level evolution may seem far from "evolution for naturalists," but understanding of its principles should help naturalists understand the complex manifestations of organic evolution that they see. For example, the principle that selection at all levels can act only by elimination, never by positive choice—a principle rarely sufficiently emphasized by evolutionists—should help naturalists understand both the visible situations that they see and the more-difficult-to-see rates and costs of evolution that profoundly affect the visible situations.

The Cosmic Theater

Many evolutionists since Robert Chambers (quoted at the beginning of my Part Two), and perhaps before, have considered the origin and evolution of life as part of cosmic evolution. This was at first only a philosophic idea, but it has now become an accepted scientific fact. A sequence of stages can be recognized which began with the origin of matter (which our limited brains may be unable to conceive of) and proceeded through the evolution of the elements (from hydrogen), our universe, our galaxy, our solar system, our earth as a whole, and the earth's surface, to the evolution of life on the earth's surface. Each of these stages has been part of or has depended on the preceding one, and the stages do therefore form a multi-level evolutionary sequence. But it is a cumulative sequence: evolution continues at all the levels (except perhaps the origin of matter), and evolutionary processes at different levels overlap and interact.

Although this sequence is accepted as a fact, the stages have been complex and are still far from completely understood. For current ideas about the origin and evolution of the elements, the universe, the solar system, the earth, and the earth's surface, see works cited in the reference list at the end of this chapter.

Evolution by Set Selection

Within the cosmic theater, evolution at all levels can be reduced to directional changes in sets. This is not an analogy but a reduction-by-synthesis of complex reality to simple, comprehensible terms.

For present purposes, *sets* may be defined as consisting of any two or more units, each a set of smaller units, that occur together or within interaction-distances, and that do or may interact. Sets that do things are *acting sets.**

Sets consist of particles of matter more or less organized at various levels, and acting sets occur at all levels of organized complexity, from sets of atoms and atom-components to sets of galaxies and universes, and from the sets of atoms that make up the simplest living molecules to the most complex sets of species. More precisely, in living systems, sets of atoms and molecules constitute genes; sets of genes, individual genotypes and population gene pools; complex sets of different units, living cells; sets of cells, multicellular individuals; sets of individuals, family groups, societies, demes, and populations; sets of these sets, species; sets of species of plants and animals, floras, faunas, and biotas; and all these together, the set that is the biosphere. All these sets, and others, form a hierarchy, but not a simple or regular one. Rather, they form an inconceivably complex whole in which evolution continues at all levels, and in which indescribably complex overlappings and interac-

---

*Although histories of ideas cannot be traced in detail here, the following item demands recognition. Baron Paul Heinrich Dietrich von Holbach (1723–1789) published his *Système de la Nature* more than 200 years ago (1770); it was translated into English by Samuel Wilkenson (3 volumes, 1820–1821); and selections from the translation are included in Goodman's *Science and Religious Belief 1600–1900* (1973, pages [271]–295). The selections include (page 273), "Every thing in the universe is in motion: the essence of matter is to act . . ."; and (page 279)". . . life . . . is in itself only an assemblage of motion." Of course some of von Holbach's other ideas were very far from being modern.

tions occur at each level and between levels. Nevertheless levels can be recognized, and it is sometimes convenient to designate a particular level as L, lower levels as $L - 1$ or $L - N$, and higher levels as $L + 1$ or $L + N$.

Characteristics of Sets and Set Selection

Directional changes in sets at all levels have common preconditions, characteristics, limiting and modifying factors, and consequences.

Preconditions. Preconditions of directional change—evolution—at all levels include *continuity,* or the continued existence of sets at least for limited times, and *variation* of sets in composition and capacity to defer disintegration, so that differential elimination and survival can occur. Another precondition of continuing directional change may be *availability of energy,* or organization of matter and energy in a universe-wide system which is low in entropy—in which energy is unevenly distributed —which allows some sets, including living ones, to draw large amounts of energy from their environments. These preconditions will be taken for granted here.

Formation and Selective Elimination of Sets. In general, sets, including living ones, are *self-organized,* by action of their members, although environmental forces sometimes'bring the members of sets together. For reading on this subject see the chapter reference list, but the references are technical and beyond the reach or needs of most naturalists.

To reduce evolution of sets by set selection at all levels to comparable terms, a definition of selection is needed that applies at all levels. For this purpose, *selection* can be defined as *differential elimination* among sets (usually but not always at the same level) which differ in their characteristics. (Biologiical definitions of selection are further considered in Chapter 6A) This is a description as well as a definition. Barring an external intelligence, which evolutionists do not detect, nothing in nature can select sets for positive preservation. The process can only be one of negative elimination in which, after different sets have formed, some are eliminated more or less promptly, while others survive for a while—that is, their elimination is deferred. The process is statistical: the elimination of a single set at a particular time may depend largely on chance, but when large numbers of sets are involved, the probability that some kinds will be eliminated while others survive may approach certainty. Of course the chances of elimination of different kinds of sets depend on their different characteristics and their relations to the particular environments in which they occur.

Elimination of sets involves their disintegration, or separation into their component parts, which are sometimes at $L - 1$ (as when individuals of a group simply separate) and sometimes at $L - N$ levels (as when a multicellular organism decomposes). And survival is always relative and temporary, at least among living sets; nothing lives forever; all living sets disintegrate sooner or later, and must do so if new sets are to be formed. All this is in accordance with the law of the conservation of matter and energy—that matter is (in this context) indestructible, although it can organize and reorganize itself in endlessly different ways.

The Pseudomathematical Equation.   Given the preceding definitions, which are also descriptions, of sets and set selection, the essential characteristics of evolutionary processes at all levels can be described by a pseudomathematical equation.

$$Mo \rightarrow SF\ (V)\ +\ DE \rightarrow DS\ =\ DC$$

Mo being movement of the units that form sets; *SF(V), formation of sets, with variation; DE,* differential elimination (disintegration); *DS,* differential survival; and *DC,* directional change.

Set Selection at Chemical Levels.   A description of directional change—evolution—by set selection as it occurs at relatively simple chemical levels may help to clarify the pseudomathematical equation. Atoms, ions, and molecules in solutions are in constant motion, as a result of their own intrinsic energy, often supplemented by energy from external sources, and their motion results in their forming various combinations, or sets. As Cairns-Smith (1971, page 66) puts it, the "molecules are buzzing about . . . trying out various arrangements." The sets are made at random (within limits in each case), and most sets are unmade—disintegrate—instantaneously. However, in appropriate environments, some kinds of sets are not unmade instantaneously; their disintegration is deferred. And it is deferment; no decision maker in the environment decides that certain combinations of molecules will be preserved regardless of circumstances; the molecules themselves, by their own characteristics, counteract destructive forces and defer disintegration for shorter or longer times. These molecules are "selected" in the sense that they continue to survive in their particular environments, and for the time being. This is natural selection at chemical levels, and it leads into and becomes natural selection at living levels.

The Energy of Evolution.   Inherent in the pseudomathematical equation are that *sets* must be formed before selection occurs, and that selection at a given level cannot directly determine what sets are formed at that level. This is an essential principle of evolution by set selection: selection—differential elimination of sets—at an *L* level *cannot* occur until processes at lower levels have made the sets. Making the sets expends energy. Sets are made either by action of their members or by environmental forces acting on the members; in both cases energy must be expended on making sets at lower levels before evolution-by-selection can occur at higher ones. This principle holds at all levels; the evolution of even the most complex living systems depends on expenditure of energy at lower levels, ultimately on the energy of nonliving atoms and molecules. This principle—that the energy of evolution comes from below—is reexamined in Chapter 6A under "The force of selection and the energy of evolution," and a consequence of it—that matter takes the forms of life because it has the inherent capacity to do so—is emphasized in Chapter 10 under "The mystery of matter."

Limiting and Modifying Factors. The essentially simple process of evolution by set selection, described by the preceding pseudomathematical equation, is limited and modified in several ways.

Competition. In theory, and sometimes also in fact, small amounts of directional change—evolution—in sets can occur without competition. Competition is therefore not essential. But, also in fact, competition usually does limit and modify evolution by set selection at all levels. At relatively simple atomic and molecular levels, one particle of matter cannot be in two places at once; when it becomes part of one set, it must often be denied to others; competition for ions is in fact a recognized factor in chemical change. At increasingly complex, living levels, competition is still not essential but in fact does limit and modify evolution in increasingly complex ways (see Chapter 6A for more detail.)

Programmed and Open-Ended Evolution. All processes of directional change are *programmed*—predetermined—to some extent by the composition of the sets concerned and by environmental factors affecting the direction, force, and limits of selection. In the case of simple chemical changes, the direction, rate, and amount or limits of change may be precisely predetermined by the kinds and numbers of atoms and molecules available to form sets (molecular combinations) and by environmental factors such as temperature, imposed pressure, and external sources of energy. In the almost infinitely more complex case of organic evolution, the direction, rate, and sometimes perhaps also the amount or limits of change are less precisely predetermined by the kinds and numbers of genes available to form sets (genotypes, which are equivalent to individuals), and by complex factors in the environment that determine the direction, force, and limits of selection.

Although partly programmed in this way, evolutionary processes are at the same time partly open-ended. From one point in evolution, change may proceed in any one of many possible directions, or in more than one, as when different treatments make different compounds from the same set of chemicals, or when groups of plants or animals radiate. In the latter case, although most of the diverging lines of evolution eventually reach limits and become extinct, some have continued through successive radiations since the origin of life and are still open-ended; their options for the future are open in many directions.

In any particular case of evolution-in-progress at any level, a ratio presumably exists between amount of programming and amount of open-endedness. This ratio may be designated P/OE. It is a significant characteristic of any evolutionary process, but one that is difficult to quantify—how can open-endedness, or the possibility of future change and diversification, be measured? However, P/OE ratios are obviously greater in some cases than in others, and cases can sometimes be arranged so that

$$P/OE_1 > P/OE_2 > P/OE_3$$

For example, $P/OE_{chem} > P/OE_{org}$, "chem" being simple chemical processes, and "org," organic evolution. This means simply that chemical processes are more precisely programmed than organic evolution is.

P/OE ratios are subject to selection, and may evolve toward either increased or decreased programming. An example is given in Chapter 5, under "Coherent populations, multicellular individuals, and ontogenies."

Feedback and Homeostasis. Feedback and homeostasis limit and modify directional change—evolution—at all levels.

Feedback occurs when the effects of a process are referred ("fed") back to the process and modify it, so that the process is accelerated (positive feedback), retarded or terminated (negative feedback), or otherwise modified by its own consequences. *Homeostasis* is the maintenance of steady states—equilibriums—in unstable systems, or of steady rates in directional processes, usually by means of alternation of positive and negative feedbacks.

A hearth fire provides a simple, familiar example—but physiologists might consider it an analogy rather than an example. Positive feedback occurs immediately after the fire is lighted, when some of the fire's heat is fed back into it and increases its rate of combustion; if this happens rapidly enough, the fire becomes an explosion. Negative feedback occurs when the fire depletes the oxygen available to it, or when inflow of oxygen is reduced by the fire's products (ashes), so that combustion is retarded. The two kinds of feedback together form a simple homeostatic mechanism that keeps the fire burning at a fairly constant rate as long as fuel is sufficient and other conditions do not change too much.

In living systems, feedback and homeostasis occur in two kinds of cases. In one, they maintain complex equilibriums, or prevent change; in the other, they allow or favor directional change—evolution—but control its rate and sometimes its direction. But these two cases are related; equilibriums are maintained by control of rates of change; in the hearth fire, an approximate temperature equilibrium is maintained by control of rate of combustion, or of change of fuel and oxygen into carbon dioxide.

Equilibrium-homeostasis is conspicuous in living organisms. For example, mammalian thermoregulation, which is essential to sustained activity and perhaps to mammalian reproduction and evolution of the mammalian brain, depends on an exceedingly complex set of positive and negative feedbacks which keep body temperature in equilibrium, within very narrow limits. The human body (like other living bodies) is in fact a set of homeostatic mechanisms, each complex in itself, and all interacting to form a literally inconceivably complex whole—complex beyond our power to comprehend it in full detail, although we can isolate details. But this is the domain of physiology and, secondarily, of medicine; for a good introduction to it see Hardy (1976).

Evolution-homeostasis is more difficult to see, because evolution itself is difficult to see, but we can usefully speculate about it. For example, if, early in the history of life, the earth was populated by free-living strands of DNA, each a set of genes, the genes being unable to live independently, selection of genes and of whole strands was presumably limited and modified by feedback. Different genes presumably varied—mutated —at different rates and in different directions, as genes do now; genes that mutated too often—were too unstable—eliminated themselves from their strands, while stable genes survived. But the instability and stability of genes were limited also by selection of whole strands. Strands of which the genes were too unstable disintegrated, and strands of which the genes were too stable did not vary, could not evolve effectively, and eventually were eliminated too. Feedback therefore favored evolution of compromise mutation rates in which gene mutations were too few to cause disintegration of most DNA strands but numerous enough to allow variation and adaptive evolution of the strands. In other words, selection at the L level (the level of whole strands) was fed back to $L - 1$ (the level of individual genes) in such a way that an effective evolution-homeostasis was maintained which favored directional change—evolution—without disintegration.

This simple example suggests three speculative comments:

1. Both genes and strands act for themselves—literally selfishly—in ways that promote their own survival; neither sacrifices itself to the other; genes that promote evolution of strands by reducing their own stability actually promote their own survival and evolution as members of surviving and evolving strands.

2. Evolutionary feedback may favor genes that mutate not only at favorable rates but also in favorable ways or directions—for example, that produce many mutants that vary only slightly rather than a few that vary drastically, or mutants that program ("code for") formation of some classes of proteins rather than others.

3. Feedback selection during evolution makes situations in which evolution occurs at favorable rates and perhaps in favorable directions; in this way evolution makes situations favorable to itself, as will be repeated in later chapters.

In summary, multi-level feedbacks and resulting homeostatic systems—both equilibrium and evolutionary ones—have been and are important, often essential mechanisms in the maintenance (deferment of disintegration) and evolution of life. They evolve in multi-level living sets when the results of selection at any level (L) are fed back to and modify actions of sets at lower levels ($L - N$) in such a way that the actions of the L sets are modified in turn. This process allows evolution by selection to make situations favorable to itself.

Consequences

The evolution of sets by set selection, as it has occurred in the universe and on the earth's surface, has consequences that seem inevitable—very strongly programmed in general directions if not in detail. One of the most conspicuous consequences is increase in diversity and organization at all levels.

Complexity, Diversity, and Organization. Several words used by evolutionists to describe characteristics of sets produced by evolution need careful discrimination.

Complexity is a characteristic of sets that consist of numerous members intricately arranged. The word emphasizes numbers and intricacy, although differences may be implied too.

Diversity is a characteristic of the members of one set that differ among themselves. The word emphasizes the differences, although it suggests numbers too. Diversity may characterize sets of which the members are separated; for example, the species of a genus exhibit diversity if they differ conspicuously among themselves even if they are geographically isolated from each other. However, when biologists speak of *tropical diversity*, they usually have in mind communities like rain forests and coral reefs, which are sets of many species of plants and animals occurring together, interacting complexly, and diversely specialized.

An *organization* is a set of which the members interact and are (usually) specialized to perform different functions, and which is capable of acting as a whole.

Organizations may be very simple. For example, a set of two stones rolling down a slope, one behind the other, meets the definition if the first stone is specialized (by its leading position) to clear a way but is stopped by occasional obstructions, and if the second stone is specialized (by its following position) to gain enough momentum in the cleared way to strike the first stone and carry both of them through the obstructions. Specialization by position occurs in many more-complex organizations, including sets of cells in developing embryos and sets of individuals in many social and cooperative situations. A water molecule, which is a set of one oxygen and two hydrogen atoms, is an organization too. From these relatively simple examples, organizations can be graded arbitrarily according to the number of members in the organized sets, number and/or degree of specializations of different members, degree of interdependence of the different members, and capacity for action of the wholes. Some of these characteristics of organizations are difficult to measure, and the different characteristics are not always well correlated. The amount or degree or organization is therefore difficult to quantify in single cases or to compare precisely in different cases. We can and do make intuitive judgments of relative degrees of organization based on the number of parts, diversity and interdependence of parts, and capacity for action of wholes. But how is capacity for action to be judged: by the number and diversity of the actions, or their precision, or their effectiveness? A water molecule acts precisely and effectively, but (comparatively speaking) not diversely. A multicellular individual plant or animal acts less precisely and effectively but more diversely. A tropical rain forest or coral reef can be considered a still more complex organization, but its capacity for action as a whole is still less precise. (If readers disagree with these assessments, the disagreement will be in a sense a measure of the difficulty of measuring organization.) Which of these sets is most highly organized? I shall not try to answer this question. But the sets do at least represent different levels of organization.

Although degree of organization is difficult to measure precisely, we can see that in general *organized diversity* (meaning what it seems to mean according to the preceding definitions) has increased cumulatively during evolution; its increase has been a conspicuous consequence of multi-level evolution in the universe and on the earth. The increase has occurred at many levels: in the evolution of the elements following the "big bang," if there was one, and in the evolutions of the physical and chemical environment of the earth's surface, of nonliving into living molecules, of living molecules into cells, of single-celled into multicellular individuals, or relatively simple groups of individuals into complexly and diversely organized insect and human societies, and of relatively simple associations into indescribably complex and diverse rain-forest and coral-reef communites. Evolution at some of these levels has sometimes included decreases as well as increases of organization or diversity; examples are the reduction of organization of many parasites, and the reduction of diversity of biotas on small islands that have become separated from continents. But evolution has produced a continual net increase of multi-level organized diversity nevertheless, with both continual increases at single levels and additions of new levels; for example, multicellular individuals and societies of such individuals can be considered two new levels of organized diversity added' during evolution.

The new levels of organized diversity have been interspersed between older levels, not added at the top of the hierarchy. Perhaps to be considered next to the top are whole biotas, which are sets of thousands of interacting species. But this is not a new level of organization. Sets corresponding to biotas must have existed from the beginning of life. At first they consisted of diverse, living or almost living, interacting molecules rather than what we now think of as organisms, but the molecular "biotas" evolved gradually and continuously, by multi-level evolution, into the most complex and diverse existing biotic communites.

The highest, most complex set in the hierarchy of life is the biosphere, which is the set consisting of all living things on the earth. It is an organization in which the members interact and are interdependent at many levels, from the broadest interactions and interdependences of green plants with other plants and animals, to the most minute interactions at lower levels. This set too represents not an added level but a product of the evolution of an ancestral set which, at the time of the origin of life, included all primitive living molecules, which presumably interacted with each other, with specific nonliving molecules, and with the earth's surface to form an evolving whole.

This concept of organic evolution as resulting in continual increase of organized diversity at many levels raises a question and poses a problem.

The question is, why has evolution resulted in continual increase of organized diversity of life on earth? The immediate answer is that there has been a selective advantage in the increase of organized diversity at living levels. We are so accustomed to this idea that we tend to take it for granted. And we can explain it by supposing that increases of organized diversity enable organisms to defer disintegration by countering increasingly numerous, diverse destructive agents in increasingly diverse environments. This is probably true at living levels, but it is not a sufficient explanation. It seems to involve circular reasoning: increase in the diversity of environments is mainly the result of increase in diversity of life, so the explanation amounts to saying that diversity of life increases because diversity of life increases. The real question seems to be, why has life at all levels—the biosphere as a whole—increased in organized diversity during evolution? A reverse process can be imagined in which extinctions and simplifications, which do occur, would outweigh increases of diversity and organization, so that the biosphere would gradually devolve rather than evolve.

A partial answer to this question is that increase in the organized diversity of life is a continuation of a universe-wide process that began with the evolution and diversification of atoms and elements, which led into an increase of diversity and organization of compounds and nonliving molecules, which in turn led into an increase of organized diversity of life at successive levels. This is perhaps a sufficient explanation for biologists. At least it shifts the onus to cosmologists.

The problem posed by the concept of evolution by multilevel set selection rises from the uniqueness of the biosphere. There is only one biosphere. Alternative, different biospheres cannot coexist with it, at least not on earth, so evolution by variation and selective elimination at this level cannot occur. A consequence may be that the biosphere as a whole evolves without the restraint that would be imposed by competition, and that its evolution is likely to lead to occasional catastrophes rather than continual adaptive adjustments.

Parts and Wholes

Whether, how, and why a whole—an organized set—is more than the sum of its parts has long intrigued philosophers. "Holism" is in fact a philosophical concept which has led some biologists beyond the limits of reductionist biology (see the references in Chapter 10). However, a reductionist holism can be derived from analysis of simple sets.

The model of two stones rolling down a slope, one behind the other, as a very simple organized whole suggests what might be called reductionist-holistic principles. (1) The Individuals in the two-stone set continue to act as individuals. (2) What each does as an individual is modified by interaction with the other individual in the set. (3) The set may act literally as a whole; the weight of the two stones together may crush through obstructions that neither could break through alone. (4) Feedback may occur; what the two stones do together modifies what each one does separately, which in turn modifies what the two do together.

The simple two-stone model provides only one pathway of interaction; each stone can influence the actions of only one other stone, although influences may be transmitted in either direction. But pathways increase in number very rapidly in more complex sets. A set of three rolling stones provides four pathways; interactions may involve stones paired in three different ways or may involve all three stones. And in still larger sets, with more members that may affect each other not only mechanically but also chemically, thermally, and electrically, possible interactions soon become inconceivably complex, beyond our power to resolve or to predict in detail.

A whole *is* more than the sum of its separate parts, but a part within a whole is less than its separate self. The whole modifies and limits what the parts do, or (in reductionist-holistic terms) the parts modify and limit each other's acts. Some philosophical consequences of this analysis are considered in Chapter 10.

Summary

Evolution occurs by set selection at many levels.

Preconditions of evolution by set selection at all levels are (1) the conservation or continuity of matter and energy (or of matter $\rightleftharpoons$ energy at atomic levels), to which is added the continuity of life at living levels; (2) the availability of energy, largely intrinsic at atomic and molecular levels, but largely extrinsic (from the sun) at living levels; and (3) perhaps a low-entropy universe in which energy is available.

Given these preconditions, the *essential process* of evolution at all levels is (1) formation of varying sets (usually by action of their members), which becomes formation of varying genotypes (individuals) at living levels, and (2) differential elimination (disintegration) of some sets and (if an evolutionary process is to continue) survival of others, which becomes Darwinian selection at living levels.

Limiting and modifying factors are (1) competition, (2) ratios of programming/open-endedness, and (3) feedback and homeostasis.

Consequences of the process include (1) directional change at all levels, which is organic evolution at living levels; (2) increase of organized diversity in the universe and in the living world; and (3) evolution of unanalyzably complex "wholes" maintained by interlevel feedback and homeostatic systems.

READING AND REFERENCES

The Cosmic Theater

Gingerich, O. et al. 1977. *Cosmology + 1*. Readings from *Scientific American*. San Francisco, California, Freeman. (Readable and well illustrated, with limited bibliographies and an index.)

Hughes, D. W. 1978. Astrotexts Updated. *Nature,* **272,** 110. (A guide to the "enormous number" of astronomy textbooks in print.)

Evolution by Set Selection

Darlington, P. J., Jr. 1972. Nonmathematical Concepts of Selection, Evolutionary Energy, and Levels of Evolution. *Proc. Nat. Acad. Sci. USA,* **69,** 1239–1243.

Goodman, D. C. (ed.). 1973. *Science and Religious Belief 1600–1900*. Dorchester, England, Henry Ling Ltd., for The Open University.

Formation and Selective Elimination of sets

Nicolis, G., and I. Prigogine. 1977. *Self-Organization in Nonequilibrium Systems* [including living systems]. New York, Wiley. (A theoretical-mathematical treatment of interest to chemists, physicists, and mathematical biologists, but beyond the comprehension of most naturalists. The authors reach the important conclusion, on page 463, that the stability of complex systems such as ecosystems and societies cannot yet be determined mathematically.)

Set Selection at Chemical Levels

Cairns-Smith, A. G. 1971. *The Life Puzzle. On Crystals and Organisms and on the Possibility of a Crystal as an Ancestor*. Edinburgh, Scotland, Oliver and Boyd.

Feedback and Homeostasis

Hardy, R. N. 1976. *Homeostasis*. Inst. of Biology's *Studies in Biology*, No. 63. London, England, Edward Arnold.

Parts and Wholes

See Chapter 10.

5

BELOW THE DARWINIAN LEVEL

This chapter is concerned with stages and components of organic evolution that could not be seen in Darwin's time or were not understood: the origin and early evolution of life; the evolution of sex, populations, programmed ontogenies, and related processes; mechanisms of heredity and variation, some of which Darwin might have seen but did not; and molecular processes that began with the beginning of life, if not before, and continue now below the whole-organism level.

These processes fit well into the concept of evolution as directional change in sets of sets interacting at different levels. This concept was hardly possible in Darwin's time. If it had been, it might have allowed prediction that the hereditary basis of plant and animal variation is particulate, and that the hereditary particles are themselves sets of smaller particles which have evolved and continue to evolve complexly at molecular as well as more-visible levels.

Origin and Early Evolution of Life

The origin and evolution of life can be considered philosophically as part of cosmic evolution (Chambers, quoted in my Part Two), or as a process of evolution of sets by multi-level set selection (Chapter 4) progressing from simple chemical to increasingly complex, organized living sets. Or the origin and early evolution of life on earth can be reconstructed as a sequence of probable events and stages based on actual evidence derived from the composition of living things and the essential chemistry of life, from analysis of ancient rocks and life products in them, from fossils, and from experiments that try to reproduce conditions that existed on the earth's surface when life originated.

Reconstructions can be less or more detailed. Bernal (1967) suggests four broad stages in the origin and early evolution of life, and Wilson et al. (1973, pages 593, 617–625), reviewing much recent work by many persons, suggest 16 "scenarios" covering the same history. For naturalists, the following seven-stage summary of probabilities and possibilities is sufficient.

1. Evolution of the Environment.   The earth originated about 4.7 billion (thousand million) years ago; core, mantle, and crust differentiated; water condensed and oceans formed, and soluble elements and compounds accumulated in them; and a "secondary reducing atmosphere" evolved consisting chiefly of water vapor, methane, ammonia, and perhaps carbon dioxide, but with no free oxygen and no layer of ozone blocking the sun's intense ultraviolet radiation. This stage of evolution of the earth and its surface made an environment fit for the beginning of life. For more details see Henderson's classic *The Fitness of the Environment* (1913: the Beacon edition, 1958, has an introduction by Wald putting Henderson into a more-recent context) and any modern book on the origin of life.

2. Subvital Molecules   Some organic (carbon-based) molecules, in solution in water, evolved increasingly complex molecular organizations. Experiments suggest that amino acids would form in the environment that probably existed then, and some of them presumably organized themselves into molecules which began to trap and use the energy of ultraviolet radiation from the sun. The molecules that did this may at first simply have accumulated radiant energy by day and released it usefully at night in ways that deferred their disintegration, but in the course of time they evolved increasingly complex ways of using energy. They became "protobionts" or "subvital molecules" at the threshhold of life.

   This stage probably began as part of stage 1 and ended (as a separate stage) at or soon after the beginning of stage 3, perhaps more than 3.5 billion years ago.

3. Replication and Lineages.   Some subvital molecules evolved the capacity to make copies of themselves—to replicate, or reproduce—using nucleic acid sequences (DNA) as templates, on which were made also diversely useful proteins. The copying was usually precise, but occasional copying errors allowed variation, selection, and evolution. The self-reproducing molecules formed lineages; efficiency of reproduction became a factor in evolution; and lineage selection was added to individual selection in further evolution. At this point life and organic evolution may be considered to have begun.

This stage cannot be dated precisely but must have preceded the formation of the oldest fossil-bearing rocks, which are more than three billion years old.

4. Photosynthesis, Oxygen Metabolism, Bacteria, and Blue-Green Algae. Living, self-reproducing, lineage-forming molecules evolved to the complexity level of bacteria. Although ultraviolet light was probably the principal initial energy source of life, several different bacteria evolved different chlorophylls by which the energy of what we now call "visible" wave lengths of sunlight began to be used to make sugars from water and carbon dioxide, with release of oxygen into the atmosphere. This was the beginning of photosynthesis. Part of the oxygen (O_2) that was added to the atmosphere was converted to ozone (O_3), which formed an "ozone shield" in the upper atmosphere which blocked much of the ultraviolet radiation that had formerly reached the earth's surface, so that ultraviolet-users were starved. This was an indirect but very effective sort of competition. At the same time different still-simple organisms evolved different ways of protecting themselves against oxygen (a strong reactive agent) and of using it advantageously in their metabolism, as all higher forms of life do.

One group of bacterialike photosynthesizers, using "chlorophyll A" (now used by all higher green plants), evolved the procaryote cell; that is they became blue-green algae, which are sometimes considered to be only specialized bacteria. This kind of cell is simpler than the later eucaryote cell, with no definite nucleus and no membrane-limited organelles, but it is nevertheless complex, with an external membrane and elaborate photosynthetic lamellae, and able to use the energy of "visible" sunlight, to use oxygen in metabolism, and (in some cases) to fix nitrogen.

Apparent fossil bacteria and blue-green algae, and apparent products of photosynthesis, have been found in rocks three billion or more years old. And bacteria and blue-greens are still highly successful. Bacteria are everywhere, and although blue-green algae do not now occur in the open sea, they are numerous in fresh water and in damp microenvironments on land, and are the algal members of some lichens.

5. The Eucaryote Cell. The Eucaryote cell evolved, characterized by a nucleus containing chromosomes, and by precise cell division—mitosis and meiosis—which is the mechanism of Mendelian heredity and variation. (The eucaryote cell is, of course, much more than this; it is an exceedingly complex organism in itself.) Its evolution permitted, or at least was followed by, more effective evolution at two new levels: of complexly organized multicellular individuals, and of coherent populations held together by gene-sharing bisexual reproduction. All higher plants and animals are eucaryotes.

The origin of the eucaryote cell—whether by a gradual evolution of one lineage of blue-green algae, or by a symbiotic, cooperative coming together of several previously independent simpler organisms—is debated. Cavalier-Smith (1975) reviews and argues for the first theory; Margulis (1970), for the second. If the origin was by cumulative symbiosis, the component organisms may have included a procaryote "amoeboid host," an aerobic bacterium (from which evolved the mitochondria concerned in oxygen-using metabolism), a spirochaetelike bacterium (from which evolved the mechanism of eucaryote cell division), and a blue-green alga (from which evolved photosynthetic plastids). See Margulis, 1970, page 58, figure 2–4.

The time of origin of the eucaryote cell is debated too. Possible single-celled eucaryote algae are fossil in rocks at least 1.3 billion years old, and Cloud et al. (1975) think they may have originated even earlier. But Knoll and Barghoorn (1975) think that the early fossils are doubtful and that the eucaryote cell probably did not appear until just before the Cambrian, less than one billion years ago, and that multicellular eucaryotes then evolved "quite rapidly." All the major eucaryote phyla first appear fossil in the Cambrian (see my Chapter 8), which strongly suggests that the kind of cell that made their evolution possible had itself evolved, or perhaps completed a long course of evolution, not long before.

6. *Multicellular Plants and Animals.*   Different single-celled eucaryotes evolved into multicellular green plants, fungi, and animals. The actual ancestors were probably flagellate protozoa, which are still numerous and which still include both many photosynthesizers and many nonphotosynthesizers, comparable respectively to the probable ancestors of higher green plants and higher animals. The first probable multicellular eucaryote plants (green algae) and invertebrate animals are fossil in rocks less than one billion years old. Additional phyla of (marine) invertebrate animals appear sparingly as fossils between then and about 0.6 billion years ago, when a more conspicuous radiation of (marine) invertebrates occurred. And the later history of multicellular plants and animals can be traced in increasing detail as the evolution of separate phyla (Chapter 8).

7. *Extension to the Land*   All the preceding stages of evolution probably occurred in or at least began in sea water, but in the course of time

organisms representing successive stages of evolution spread to the land. This was not evolution to a new level, but an extension of the theater of evolution. Photosynthesizing bacteria and blue-green algae may have occupied damp microenvironments on land very early in their history, and appropriate protozoa probably joined them in due course. Then, some eucaryote algae probably spread into fresh water if not actually to land well before the appearancce of true land plants. The first multicellular land plants, perhaps derived from fresh-water green algae, are fossil in the late Silurian more than 400 million years ago, and primitive land plants diversified conspicuously during the following 25 million years of the Devonian. Animals can have occupied the land only after plants had done so. Among the first were probably various worms and other invertebrates most of which were soft-bodied and have left no fossil record of their arrival on land. Major "invasions" of land have been made also by gastropod mollusks (snails), by several separate groups of arthropods, and by vertebrates, which came onto land more than 300 million years ago. Additional, minor invasions and almost invasions have been made by other animals, and are still going on. Some marine snails now extend into damp places above tide lines. Some existing shrimps and crabs have moved onto land, especially in the tropics. Even some fishes go onto the edge of land: mud skippers, climbing perches, an oriental catfish, and even common eels, which can move short distances over land through wet grass.

Many, but not all, of these animals, including the first land-living vertebrates, probably reached the land by way of fresh water rather than directly from the sea. In these cases, adaptations to seasonal fresh-water habitats were probably sometimes pre-adaptations to life on land. The pre-adaptations probably included ability to breathe air when water dried up, and ability to disperse across land from one isolated pool or stream to another. Dispersal may have been the principal, initial, selectively advantageous function both of crawling on land by the ancestor of land-living vertebrates and of flight by the first winged insects. For further details see Chapter 8.

The Process as a Whole   The details suggested in the preceding seven-stage history can be only an infinitesimal part of the actual process of origin and evolution of life on earth. The process must have been a continuity, with complex transitions from stage to stage, and with the stages overlapping complexly. It probably involved continual fanlike radiations, with competition, selective elimination, and differential survival, followed by reradiations (Chapter 6C). This has been the pattern of evolution of the higher forms of life of which we know the histories

(Chapter 8), and it is likely to have been the pattern from the beginning. Protobionts and the earliest living molecules probably diversified, competed, and underwent selective eliminations; the threshhold of life may have been approached or crossed many times, although apparently only one crosser survived and evolved to higher levels. And this process of diversification and selective elimination probably occurred innumerable times at successive sublevels and levels during the three billion years or so of the origin and early evolution of life. The process was a continuity, but an exceedingly complex one—an inconceivably complex cumulative sequence that took more than three thousand million years to form the layered whole that is the existing biosphere.

Alternatives and Difficulties   Bernal (1967) emphasized the difficulties, complexities, and doubts involved in attempts to reconstruct the actual history of the origin and early evolution of life: his Chapter 8 is concerned with "alternative pathways," and his Chapter 9, with "outstanding difficulties." Alternatives include the possibility that the precursors of living molecules were crystals (Cairns-Smith, 1971), or that living molecules reached the earth from elsewhere in or outside of the universe (Hoyle and Wickramasinghe, 1978). However, dramatic alternatives like these seem unnecessary and unlikely. A history something like the one outlined here is supported by evidence from many sources, especially by the survival of an essentially complete series of organisms which, although they cannot be actual ancestors, do seem to represent the stages of an evolutionary process that occurred *on the earth*.

Evolution of Mixis, Sex, Populations, and Ontogenies

The evolution of mixis, sex, coherent populations and programmed ontogenies can be considered a multi-level continuity. It will be described as such here, with stress on simple concepts but reference also to actual complexities, and with some speculations added to the facts.

Mixis.   Before and during the origin of life, dissociation of parts of complex organic molecules and self-assembling of parts into new, variable molecules presumably occurred. This was the beginning of "mixis," or of the continual rearrangement of hereditary particles essential to evolution. It probably made early phylogenies reticulate, with different lineages separating and then joining again, forming networks rather than successive fans.

Mixis continued during the early evolution of life but was for a long time irregular, and some phylogenies probably continued for a while to be reticulate. Mixis is still occasional and irregular among bacteria.

Bisexual Reproduction. In multicellular eucaryotes mixis has evolved into bisexual reproduction, by coevolution of two sexes which are differently specialized, males to fertilize females, females to produce offspring. But this simple principle is diversely manifested. In flowering plants, male and female flowers are in some cases separate, borne either on the same or on different individual plants, but in other cases the sexes are combined in the same "perfect" flowers. In multicellular animals the sexes are usually separate, but are combined (in the same individuals) in exceptional cases.

The differentiation of sexes in multicellular animals is genetically determined, but the genetic mechanisms are different in different groups. In human beings sex is determined by a xx pair of chromosomes in females and an xy pair in males; in some insects, by whether a particular chromosome is single or double; and in ants and bees, by whether the eggs are fertilized or not, which is to say by whether individuals are diploid (females) or haploid (males).

This diversity suggests that differentiation of sexes is not inherent but that the different mechanisms of it have evolved in different cases because separation of the sexes is selectively advantageous. The primary advantage of this is presumably that it enhances mixis, a presumption reinforced by the fact that many plants and animals have secondary adaptations that favor outcrossing, or mixing of the genes of not-too-closely-related individuals.

That bisexual reproduction has these advantages is an hypothesis. Two serious questions can be asked about it. One is, how can an individual afford to give up half its fitness—which it is said to do when it gives its offspring only half its genes, the other half coming from another individual—without receiving a more-than-compensating selective advantage? Williams (1975, page 7, citing Bonner, Maynard Smith, and others) calls this "the paradox of sexuality." The answer probably is that, although they seem to give up half their fitnesses, individuals of both sexes gain more-than-compensating advantages that actually increase their fitnesses—that is, increase the chances that their genes will be represented in later generations.

The other question is, why are the sexes usually in separate individuals, at least in animals? Evolution can produce individuals—hermaphrodites—that both lay their own eggs and fertilize the eggs of other individuals, so that two individuals can fertilize each other and both can then lay eggs, as slugs and earthworms do, and this should double the efficiency of reproduction, other things being equal. The answer is, presumably, that other things are not equal, that separation allows separate specializations which increase the efficiency of both sexes and, consequently, increase the efficiency of their cooperative reproduction.

Differentiation of two sexes allows not only their reproductive specializations but also further differentiation of roles. For example, if males and females differ in size, they may eat partly different foods and so may make more effective use of diverse resources. Or if males are larger, as in many but not all vertebrates, they may be specialized to defend the smaller females and offspring.

For further reading on the evolution and complexities of sex, see the books cited in the chapter reference list.

Coherent Populations, Multicellular Individuals, and Ontogenies.   The eucaryote cell, which evolved perhaps one billion years ago, can live independently; single-celled eucaryote "protists'" are conspicuously successful. But eucaryote cells are also able to form sets in ways that procaryotes apparently cannot do.

Sets of eucaryote cells did evolve in two divergent directions. In one direction they formed populations that were more coherent and more complexly organized than those of procaryotes. Within each population, lineages evolved as before, but the lineages were held together by obligatory (rather than occasional) bisexual reproduction, with Mendelian exchange of genes, so that phylogenies became again reticulate in a limited and orderly way. Lineage as well as individual selection occurred within populations. Lineages that were not flexible and adaptable were eliminated, while open-ended lineages survived and evolved in directions and at rates determined by selection. Of course lineage selection and evolution were not new; what was new was the way in which eucaryotes formed *coherent* populations that were notably flexible and adaptable.

In the other direction, within some lineages, some cells began to form themselves into multicellular sets, by replication of single cells without separation of the "daughters." These sets were probably at first loose colonies of undifferentiated cells, but during the course of evolution they gradually became increasingly organized and programmed. Different cells in a set became specialized for different functions; reproduction of new sets became limited to cells not specialized for other functions; and the sets began to evolve specific patterns of development, or ontogenies, which can be thought of as short-term segments of evolution programmed both in details of development and in termination of the sets' existence. These sets became what we now call "individual" multicellular plants and animals, which develop in predetermined ways and die within predetermined times.

In short, what we now distinguish as phyletic evolution in populations and as individual ontogenies have both evolved by selection of sets of eucaryote cells, but selection has proceeded divergently, so that now, conspicuously,

$$P/OE_{phylogeny} < P/OE_{ontogeny}$$

The relation between single reproductive cells and the multicellular individuals they produce is noteworthy. The reproductive (or germ) cells form the primary set in a lineage or population; they exchange parts (genes), vary, compete, and evolve by selection in the same way that free-living cells do. The multicellular individuals are secondary sets, constructed (programmed) by the reproductive cells as means of surviving and competing. Dawkins calls them "survival machines." The multicellular individuals can be thought of also as trial vehicles, made by the reproductive cells in an endless series of trials of ways of surviving in continually changing environments. Some trials are successful, some are not; this is natural selection.

From this point of view, the question of which came first, the egg or the chicken, is easy to answer. The egg, which is a single reproductive cell programmed to make a chicken and containing the materials to do it, came first in evolution, and the chicken came second. The egg uses the chicken to make another egg, as Samuel Butler said in different words.

Tri-determinant Programming. The evolution of multicellular individuals required an increase of genetic programming. The reproductive cells had to carry not only genes that programmed their own development, reproduction, and behavior, but also additional genes that did three things: (1) kept the cells of a set together (inhibited their dispersal), (2) made all the cells susceptible to differentiation according to their positions in a set and to cell interactions; and (3) programmed an increasing diversity of special behaviors of cells in special situations within a set. This was a "tri-determinant" system (Darlington 1972b). Note that all (or almost all) the cells in a multicellular individual organism receive the same set of genes, and that differentiation of cells in ontogeny is therefore not genetic but environmental; differentiation of cells during ontogeny is in fact by something like "direct induction" or "inheritance of acquired characters," as noted later in this chapter. Something more will be said about tri-determinant gene systems in Chapter 7, under "Social groups."

Highly organized multicellular organisms are all eucaryotes. Procaryotes existed for two billion years before eucaryotes, but apparently never produced comparable multicellular forms. Perhaps their genes cannot organize themselves into effective tri-determinant systems.

Termination of Ontogenies.   During the evolution of ontogenies, feedback selection presumably favored lineages in which the life spans of individuals were long enough to permit effective reproduction, but not too long; if individuals had lived too long, they would have competed with later generations and retarded evolution. The removal of older individuals actually increased the representation of their genes in later generations. Life spans need not have been shortened by special mutations. Most mutations are deleterious, and selection need only have allowed enough such mutations to accumulate to adjust ontogenies to selectively advantageous average lengths. Environmental factors such as age-related diseases might have had the same effect, and if so, they were advantageous and may have coevolved with their hosts. But some biologists may consider this idea—that termination of life spans may be selectively advantageous, even if the termination is caused by disease— too highly speculative.

The Biogenetic Law.   Haeckel's "biogenetic law," that ontogeny recapitulates phylogeny, was once accepted as an obvious truth, then abandoned by most biologists, not because it was wrong in principle, but because some of Haeckel's details were wrong, and because most biologists were impressed by the wrong details more than by the simple principles. In simple principle ontogeny does recapitulate phylogeny, but the recapitulation is at best incomplete and often also imprecise and complexly modified by omission of stages, distortion of sequences, or premature termination, as perhaps in the origin of chordates (Chapter 8). Nevertheless, ontogenies of multicellular individuals do begin with single cells (except in some kinds of asexual reproduction) and often include striking ancestral details like the gill-clefts in the human embryo. See Gould (1977) for a perhaps too critical discussion of the concept of recapitulation.

Mendelian Heredity

In conventional Mendelian genetics, a *gene* is an hereditary unit occurring at a fixed locus on a chromosome. Genes are now known actually to be complex sets of smaller units that are difficult to limit and define precisely; some of the complexities and difficulties are described in a readable way by Dawkins (1976, pages 30–35); nevertheless, genes do act as units in simple Mendelian processes. An *allele* or *allelomorph* is one of two or more alternative forms of a gene. An allele *is* a gene; calling it an allele indicates only that different forms of the same gene, at the same locus, occur in different individuals.

Mendelian heredity—the mechanism of it—evolved as part of the evolution of the eucaryote cell. It is the principal mode of heredity of sexually reproducing plants and animals—the means by which their genes make new sets of themselves. Its essential characteristics are that it is particulate (as all heredity is at one level or another), and that the particles—genes or alleles that (interacting with the environment) determine the hereditary characteristics of individuals—are inherited according to "laws" first discovered by Mendel (Chapter 1). Mendel actually discovered, not the genes themselves, but their visible effects and statistical relationships, including the suppression of the effects of "recessive" alleles in the presence of "dominant" ones in the F_1 (the first generation after a cross), and the 3:1 and other simple ratios in the F_2. Simple Mendelian heredity is complicated by multiple alleles, pleiotropic genes, polyploidy, linkage, crossing over, incomplete dominance, and so on almost *ad infinitum;* and modern geneticists emphasize the integration of gene sets as wholes and the roles played by "regulatory" and "switch" genes. And the addition of molecular to classical Mendelian genetics has still further increased the complexity of the subject, although at the same time illuminating it. All this is well covered in many recent books; to treat it in detail here would be a waste of space.

Genetic Dominance and the Allele-Equilibrium Principle. However, two facts should be emphasized for naturalists who are not geneticists. One is that Mendelian "dominance" confers no selective advantage; "dominant" alleles hide but do not eliminate recessive ones and do not increase in a population because they are dominant. The second fact, which follows from the first, is that the genetic composition of a population—the proportions of different genes or alleles—is in equilibrium— stays the same from generation to generation—except for fluctuations due to chance, unless the proportions are changed by selection (or by mutation or by individuals entering or leaving the population). This is usually called the Hardy-Weinberg law, but the self-explanatory name *allele-equilibrium law* is preferable. Consequences of this "law" are that large populations tend to maintain their genetic diversity and may increase it if selection acts on a population in more than one direction, while small populations tend to lose alleles by chance and thus to change by "genetic drift" (Chapter 6A).

How Genes Make New Sets of Themselves   A principle of evolution by multi-level set selection is that new sets are (usually) made by action of their members. At first thought this seems not to be the case in Mendelian heredity, in which genes are "shuffled" passively during reduction division and recombination. But the mechanism that shuffles them—that pulls them apart and puts them together again—is made and programmed by the genes, which thus do act, if indirectly, to make new sets of themselves.

Non-Mendelian Heredity

Heredity does not always follow Mendel's laws. In special cases it is determined by genes in plastids outside the nucleus in eucaryote cells, or by the cytoplasm; in these cases heredity is only through the female, by way of the egg, for the male sperm carries no plastids and little cytoplasm.

In some other cases heredity involves something like *direct induction* and/or *inheritance of acquired characters.* Literally, these processes occur whenever the environment induces a heritable change in an organism, as when radiation causes a heritable mutation, but when evolutionists speak of "inheritance of acquired characters," they usually refer to the theory that *useful, adaptive* changes induced in individuals during their lifetimes are inherited. The trite example is the giraffe's neck, which (according to the theory) becomes longer as the individual giraffe stretches it to reach high foliage, the longer neck then being transmitted to the giraffe's offspring. This theory is usually attributed primarily to Lamarck, but unfairly so (Chapter 1). It is rejected by most modern evolutionists because no clear case of it and no mechanism for it have been found in the eucaryote plants and animals that most evolutionists study. In fact it was rejected by some pre-Darwinians including Prichard who said, probably in 1813, "All acquired conditions of the body end with the life of the individual in whom they are produced. But for this salutary law the universe would be filled with monstrous shapes." I shall repeat this telling phrase in Chapter 10. I have it, not from Prichard, but as quoted by Leakey and Lewin (1977, page 28).

However, something like "inheritance of acquired characters" does occur at other levels of multi-level evolution. It occurs at chemical levels: complex molecules often evolve by stages, acquiring new characters at each stage in ways determined by the environment, and carrying them into succeeding stages. This process presumably continued into the origin and early evolution of life, as long as (if) the living molecules consisted entirely of DNA (or comparable molecules); changes directly induced in the DNA by the environment were transmitted during later replications. But as soon as living molecules began to insulate themselves from the environment, by means of protein sheaths or in other ways, direct induction and inheritance of acquired characters decreased, and became negligible in *phyletic* evolution at the level of eucaryotes.

Nevertheless, even among the most complex eucaryotes—higher plants and animals—"inheritance of acquired characters" may be approximated or simulated. If segments of DNA can be deleted from or inserted into the chromosomes of an individual's reproductive cells by genetic engineering, the changes will be directly hereditary; and changes in cytoplasm induced by the environment may be transmitted directly too, from females to offspring. And it is a conjecture consistent with the multi-level point of view that, although the environment cannot induce favorable mutations directly, it may, by multi-level feedback selection, have some effect on the classes of mutations that occur and may, within limits, improve the ratio of constructive to destructive ones. (See Chapter 4, under *"Feedback and homeostasis."*) This, if it happens, is not direct induction, but the next thing to it. But this may be pushing speculation a little far.

Although direct induction and inheritance of acquired characters became negligible in *phyletic* evolution among higher plants and animals, they are conspicuous in special cases. The ontogenies of multicellular individuals can be thought of as strongly programmed segments of evolution by inheritance of acquired characters, different cells acquiring different characters in their different environments among other cells, and transmitting the characters to their daughter cells. Thinking—chains of thought, each one leading to the next—is a comparable process. And cultural evolution too involves inheritance of acquired characters; behaviors and ideas acquired by individuals during their lifetimes are transmitted to other individuals (Chapter 10).

To summarize: from the multi-level point of view, direct induction and inheritance of acquired characters occurred at chemical levels before the origin of life, perhaps continued into the early evolution of life, then diminished during phyletic evolution and became negligible in the phylogenies of eucaryotes, but became important again in ontogenies of multicellular eucaryotes, in logical processes in single brains, and in cultural evolution in sets of brains.

To turn from the multi-level to the naturalist's point of view: the visible adaptations of plants and animals may seem endless in their variety and perfection (although they may often be less perfect than they seem), and the simplest explanation of their evolution may seem to be that each individual adapts itself to its environment, and that the adaptations are directly transmitted from generation to generation, as if evolution were a simple walk in a favorable direction. Individual organisms that enter new environments can often be seen to be directly modified there by environmental factors such as cold, or wind, or limitation of resources. And subpopulations that have been established in new environments can often be shown, later, to be genetically differentiated. How is the naturalist to know that the adaptive modifications have evolved, not by inheritance of acquired characters, but by random variation and selection in the new environment? It is important that naturalists do understand this, that they understand that adaptations do not evolve by simple "walks" but by complex, random walks of which the direction is determined by selective eliminations of enormous numbers of individuals that walk in the wrong directions, and which is a slow, complex, costly, trial-and-error process that in many cases is never completed. This has been the pattern of evolution of man as well as of other organisms.

Mutation

From a multi-level point of view, mutations occur in sets at all levels, from the sets of atoms that make up molecules to the sets of species that make up biotas, whenever a member of a set is changed, lost, or added. Evolutionists usually think of mutations only in hereditary material, but recognize gene mutations, chromosome mutations, and also somatic mutations, which are genetic changes in somatic cells which change the course of ontogenies, producing, for example, a new kind of apple on one branch of a tree, or a cancer rather than a normal organ in an animal.

The causes of gene and chromosome mutations are partly inherent instabilities and partly environmental. Mutations can be caused—or, rather, their probability increased—by temperature, many chemicals, natural and man-made radiations, and other genes; some genes do modify the mutation rates of others.

Mutation has both good and bad effects. It is essential to evolution. Genetic systems that did not mutate could not evolve. But the direction of mutation is random (within limits), not determined by the needs of evolving organisms. Most mutations are in fact injurious and are eliminated from populations only by selection at heavy cost (Chapter 6B). Actual mutation rates are compromises; mutations must occur in numbers that allow effective evolution, but do not destroy or handicap too many individuals. Such compromise mutation rates are determined by multi-level feedback selection (Chapter 4).

Human evolution must have involved this compromise: mutations must have occurred in numbers that permitted rapid and effective evolution, but did not cost the loss of too many individuals. However, although the cost was bearable, it was and still is heavy in terms of individuals who die prematurely or are handicapped by injurious mutant genes. It seems likely that a mutation rate that allowed the evolution of an ape into man in a few million years, and that we have inherited, is much higher than is needed or comfortable now, but we cannot easily decrease it and must continue to pay for it. In fact we are probably increasing rather than decreasing mutation rates in human populations by increasing the amounts of chemical mutagens and of radiation in our environment. The first known environmental carcinogen was chimney soot from wood and coal fires. It would be interesting to know how soot in the environment compares with man-made radiation as a cause of cancer now!

Molecular Evolution

Ayala says (1976, page VII):

> The systematic application of the conceptual models and techniques of molecular biology has provided new insights into evolutionary processes, and new methods in the reconstruction of evolutionary history. Techniques such as gel electrophoresis, immunodiffusion, protein sequencing, and DNA hybridization have become powerful means to study biological evolution. Indeed, the last 10 years have witnessed the blossoming of a new field of study whose subject is biological evolution at the macromolecular level. This new biological discipline might be called 'Molecular Evolution'."

Although this kind of evolution is out of sight below the visible Darwinian level, even laymen know something of the spectacular accomplishments and potentialities of it in gene-splicing (making recombinant DNA) and genetic engineering. However, molecular evolution cannot be treated in much detail here. The study of it requires special techniques, including those named by Ayala, and the significance of some of the findings is still doubtful.

Principles of Molecular Evolution. The principles of molecular evolution as seen by a multi-level evolutionist and naturalist (other biologists may see them differently) are these:

1. Evolution continues at molecular levels within, and partly independent of, evolutionary processes at higher levels. This is expected from the theory of multi-level evolution.

2. At molecular as at higher levels, variation is particulate. Acting sets of atoms and molecules form themselves, vary, sometimes compete, and evolve by a process of differential elimination (disintegration) and differential survival.

3. At molecular levels, variation and heredity are partly not Mendelian but more simple and direct. Molecules evolve partly by direct induction—something like inheritance of acquired characters (see earlier in the present chapter). Changes in complex macromolecules, presumably partly induced by the molecules' immediate environments, are inherited by the molecules' "descendants." And some molecular lineages therefore evolve and diversify relatively rapidly and with relatively little restraint before selection occurs.

4. The direction of their evolution is determined partly by selection and partly by random or "neutral" processes; how much of molecular evolution is selective and how much selectively neutral is not yet known.

5. Reproduction of living organisms is by replication at molecular levels; usually by separation of the complementary strands of two-stranded DNA and formation of a new complementary strand on each of the old strands, which act as templates. Repeated molecular replication—production of surplus individuals—is the source of the energy of Darwinian evolution (Chapter 6A). And "copying errors"—molecular mutations—during this process produce heritable variations which allow selection and evolution at the Darwinian level.

6. Molecular sets of DNA—genes—determine the development of individual organisms by transcription, translation, and formation and interaction of diverse proteins. This process determines ontogenies, which are comparable to programmed segments of evolution. The developmental process is exceedingly complex, and is now under intense and fruitful study.

The Selectionist-Neutralist Controversy. Molecular-evolutionary studies are now dominated by the selectionist-neutralist controversy. Protein (enzyme) polymorphisms have been found (by electrophoresis) to be unexpectedly numerous in probably all natural populations. The controversy is about whether slightly different isozymes, which are molecular variants of what were once thought to be single proteins, and which are produced by slightly different forms of what were once thought to be single genes, differ in their biological actions, and whether they evolve by selection or are selectively neutral.

This controversy cannot be followed in detail here, but a speculative comment can be made about it. Most evolutionists are convinced "selectionists," who have been educated to believe that selection precisely determines all the characteristics of organisms at the Darwinian level, and these persons may automatically extend the belief to the molecular level—as I used to do myself. But (1) if Darwinian evolution includes nonadaptive genetic drift, and if Darwinian selection is so slow and costly that adaptations are usually imperfect (Chapter 6B), with many details not determined by selection but (in effect) neutral, and (2) if direct induction magnifies nonselective variation at the molecular level, then "neutrality" in molecular evolution is brought into better perspective: recognition of considerable neutrality at the Darwinian level makes even more of it in molecular evolution acceptable to naturalists like myself.

Speculations like this are justified, I think, in a field in which sophisticated analyses have been inconclusive and can "explain anything" (Soulé, in Ayala 1976, page 76).

The Genetic Code and Programming. Molecular biologists speak of "the genetic code," made up "base triplets of DNA and RNA," which are compared to the letters of an alphabet. Heredity and development are "specified" by sequences of the "letters" on strands of DNA. The letters and sequences are said to carry "coded information," which is compared to the information carried by sequences of letters in a printed message. And the genetic "message" is said to be "read" by evolving and developing organisms. This is an effective explanatory and teaching analogy, but it can be misleading. It is too useful and too entrenched to be discarded, but another, closer analogy, which is actually a description, may help to clarify it.

Genetic material is made up of sets of nucleotides, triplets, genes, and sets of genes that act. Even the nucleotides of a DNA "template" act, first by "holding hands" with complementary nucleotides and then by letting go. Words in a printed message do nothing like this. And the acting sets are programmed. During the development of an individual eucaryote, a set of genes programs the formation and the acts of a set of cells, this set programs the next set of cells, and so on. And during evolution, sets of genes not only program development of successive individuals but also program replication of themselves, the programming being (usually gradually) modified by mutations. This analogy-description treats genetic material, not as inert like a printed message, but as the living material it is, and makes the "genetic code" a code of molecular behavior.

Summary

This chapter is itself only a summary of probable stages in the evolution of life on earth, and of the place in it of mixis and sexual reproduction, populations and ontogenies, Mendelian and non-Mendelian heredity, and continuing molecular processes.

Readers may find it useful to compare this chapter with C. D. Darlington's (no relation) "A Diagram of Evolution" (1978), which summarizes a chromosome-geneticist's view of evolution below the Darwinian level. Darlington's diagram focuses on complexities, the present book on simple principles, but some conclusions are the same. Both treat evolution as a multi-level continuity, with processes at different levels continuing partly independently as well as interacting with feedback. And Darlington's "forward looking . . . capacity of evolution" seems roughly equivalent to the principle, stated repeatedly in the present book, that evolution makes situations favorable to itself, by multi-level feedback.

READING AND REFERENCES

Origin and Early Evolution of Life

Bernal, J. D. 1967. *The Origin of Life*. London, England, Weidenfeld and Nicolson; Cleveland, Ohio, World. [Includes reprints of historically important papers by Oparin (1924) and Haldane (1929).]

Wilson, E. O., et al. 1973. *Life on Earth*. (*See* my Chapter 3.)

Evolution of the Environment

Henderson, L. J. 1913. *The Fitness of the Environment*. New York, Macmillan. (A classic; reprinted 1958, Boston, Massachusetts, Beacon Press, with introduction by George Wald.)

Various authors. 1975. *The Solar System. Scientific American*, **233,** September. (Includes articles on the evolution of the solar system and of the earth.)

The Eucaryote Cell

Cavalier-Smith, T. 1975. The Origin of Nuclei and of Eukaryote Cells. *Nature,* **256,** 463–468.

Margulis, L. 1970. *Origin of Eukaryotic Cells.* New Haven, Connecticut, Yale Univ. Press.

Cloud, P., M. Moorman, and D. Pierce. 1975. . . . Late Proterozoic Cyanophyte; Some Implications for . . . Plant Phylogeny. *Quart. Rev. Biol.,* **50,** 131–150.

Knoll, A. H., and E. S. Barghoorn. 1975. Precambrian Eukaryotic Organisms: A Reassessment of the Evidence. Science, **190,** 52–54.

Extension to the Land

Goin, C. J., and O. B. 1974. *Journey onto Land.* New York, Macmillan; London, England, Collier Macmillan.

Banks, H. P. 1970. (*See* Chapter 8.)

Romer, A. S. 1968. (*See* Chapter 8.)

Alternatives and Difficulties

Cairns-Smith, A. G. 1971. *The Life Puzzle. . . . The Possibility of a Crystal as an Ancestor.* Edinburgh, Scotland, Oliver and Boyd.

Hoyle, F., and N. C. Wickramasinghe. 1978. *Lifecloud: The Origin of Life in the Universe.* London, Dent; New York, Harper & Row.

Evolution of Mixis, Sex, Populations, and Ontogenies

Darlington, P. J., Jr. 1972a. Nonmathematical Concepts of Selection, Evolutionary Energy, and Levels of Evolution. *Proc. Nat. Acad. Sci. USA,* **69,** 1239–1243.

Bisexual Reproduction

Ghiselin, M. T. 1974. *The Economy of Nature and the Evolution of Sex.* Berkeley, California, Univ. California Press.

Williams, G. C. 1975. *Sex and Evolution.* Princeton, New Jersey, Princeton Univ. Press.

Smith, J. M. 1978. *The Evolution of Sex.* Cambridge, England, and New York, Cambridge Univ. Press. (Probably too theoretical-mathematical for most naturalists.)

Tri-determinant Programming

Darlington, P. J., Jr. 1972b. Nonmathematical Models for Evolution of Altruism, and for Group Selection. *Proc. Nat. Acad. Sci. USA,* **69,** 293–297.

The Biogenetic Law

Gould, S. J. 1977. *Ontogeny and Phylogeny.* Cambridge, Massachusetts, Harvard Univ. Press.

Mendelian Heredity

Dawkins, R. 1976. *The Selfish Gene.* Oxford, England, and New York, Oxford Univ. Press.

Beadle, G., and M. Beadle. 1966. *The Language of Life, an Introduction to the Science of Genetics.* New York, Doubleday, Anchor Books ed., 1967. (Possibly still the best introduction to modern genetics for the intelligent layman.)

Spiess, E. B. 1977. *Genes in Populations.* New York, Wiley. (Includes a more technical introduction to conventional genetics.)

Hardy, G. H.; Weinberg, W. 1908. [Papers published independently; *see* Dunn, 1965, *Short History of Genetics,* for references. These were the first general mathematical statements of the allele-equilibrium principle, which is essential to understanding evolution, but the principle was "implicit" in earlier papers by Mendel and Pearson (Dunn, page 122).]

Non-Mendelian Heredity

Beale, G., and J. Knowles. 1978. *Extranuclear Genetics.* London, England, Edward Arnold.

Leakey, R. E., and R. Lewin. 1977. (*See* my Chapter 9.)

Lyman, H. 1979. Chloroplasts and Mitochondria: the Search for a Grand Overview. *Quart. Rev. Biol.,* **54,** 165–168. (A review of recent literature on the biochemistry, genetics, and evolution of organelles outside the nucleus.)

Mutation

Auerbach, C. 1976. *Mutation Research. Problems, Results and Perspectives.* New York, Halstead (Wiley). ["A readable and useful book" (*BioScience*); technical, but persons who wish to make correct decisions about mutagens and the risks of mutation *must* have technical information.]

Molecular Evolution

Ayala, F. J. (ed.). 1976. *Molecular Evolution.* Sunderland, Massachusetts, Sinauer Associates.

Darlington, C. D. 1978. A Diagram of Evolution. *Nature,* **276,** 447–452.

6

EVOLUTION AT THE
DARWINIAN LEVEL

This chapter deals with evolution at the visible Darwinian level, or with descent with modification of the plants and animals we see around us, and of ourselves. This is what most people mean by "evolution" and what most books on evolution are about, and what naturalists see.

For the convenience of readers, this long chapter is divided into three sections. Section A treats prerequisites and components of Darwinian evolution, with emphasis on competition and selection; B, characteristics of the process as a whole, including its inherent patterns, rates and modes, costs, and how costs limit adaptations; and C, consequences of the process, especially adaptation, diversification, and extinction, and this section ends with familiar but thought-provoking examples of evolutionary situations that naturalists can see.

SECTION A. PREREQUISITES AND COMPONENTS
Individuals and Populations

Evolution at the Darwinian level is essentially directional change in sequences of individuals. The individual is always the unit of Darwinian selection; even "group selection" can be translated into terms of selection of individuals in environments that include other individuals (Chapter 7). However, individuals usually form genetically bonded populations (Chapter 5), which are so important that most work on evolution is concerned with them.

The individual and population points of view are not antagonistic. Individuals are samples of their populations, more limited but not otherwise different from samples collected by population biologists. Modern

Darwinian evolutionists as well as systematists are very conscious of the fact that the individual is a sample. No biologist who recognizes this fact is a typologist. Any biologist who does not recognize it, but thinks of individuals as idealized types or of populations as idealized mathematical abstractions, is a typologist in the true, Greek sense (Chapter 1). (Taxonomists' "types" are not idealizations but actual specimens designated and preserved for reference.)

Prerequisites and Components

Evolution at the Darwinian level has five prerequisites or components: continuity of life, multiplication of individuals, variation, competition, and selection. In theory small amounts of evolution can occur without some of these components, but in fact they have all been essential to evolution as it has actually occurred.

The Continuity of Life

The continuity of the more visible forms of life had been accepted by almost everyone, even by creationists, as a practical matter before Darwin's time, but it was only in his time that continuity of life at microscopic levels was established, after long controversy (Farley, 1977). Now, of course, biologists accept it without question. When life originated, living molecules and sets of molecules of DNA added their continuity as organized, replicating sets to the continuity of matter and energy. And since then life has come only from preexisting life.

Multiplication

Multiplication, or increase in numbers of individuals by excess reproduction, is essential to evolution by selection; continual selective eliminations require continual production of more than enough individuals to maintain a population. We now know the basis of it, which is *repeated* replication at molecular levels (of DNA or RNA), but recognition of the fact of it was all that Darwin and Wallace needed, and all that naturalists need now, to understand evolution at the Darwinian level.

Variation

Variation occurs among all living things. Some of it is genetically determined and hereditary; some, not hereditary but environmental, caused by variations in temperature, moisture, food, and so on. But note: all the

characteristics of individuals are produced by (indirect) interactions between the genes and the environment during ontogeny; all therefore
have a genetic base, which makes the difference between, for example,
a man and an ape; but the product of genes × environment may be varied by variation of either the genes (hereditary variation) or the environment (environmental variation).

Only genetically determined, hereditary variation is directly concerned in Darwinian evolution. Its occurrence, measurement, and
"uses" cannot be described here; they are covered in genetics texts. The
subject is endlessly complex. Wright treats it in his four-volume *Evolution and the Genetics of Populations* (1968–1978), of which the last
volume is subtitled *Variability within and among Populations.* Lewontin
says that the subject of his 346-page book, *The Genetic Basis of Evolutionary Change* (1974), is "the nature of genetic diversity among organisms." But naturalists cannot use books like these. For the most part
they have to make do with the fact of variation, which they can see as
Darwin did, and with a bare outline of the underlying principles of
Mendelian heredity and, underlying that, molecular processes.

Competition

Competition occurs among organized sets at all levels; matter and energy taken by one set must often be denied to others (Chapter 4). Now
to be considered are its characteristics at the familiar, Darwinian level.
Darwin emphasized its importance in Chapter 3 of the *Origin.*

In ordinary usage, *competition* means a contending of rivals for almost anything, either directly by almost any means ("all is fair in love
and war") or indirectly (as for resources or markets in commerce). Naturalists' conceptions of competition are often equally broad, but theoretical-mathematical biologists either do not define it at all or define it
more narrowly, as including only "actions and interactions," or "ex-
ploitations and interferences," or "active demand" of individuals or species seeking the same resources. This definition is intended to be
precise, but in fact is not. It does not specify that the resources competed for must be not only limited (as all resources are) but actually
limiting. It does not divide "active demand" into noncompetitive and
competitive fractions. And it does not define "resources." Food and (for
green plants) light clearly are resources which are competed for. But
mates, and positions in peck orders, are competed for too. Are they resources?

Biologists usually exclude *predation* from their definition of competition, and this raises further questions. When a fox kills and eats a rabbit,

is this competition, and if not, how is it related to competition? Foxes and rabbits seem to compete for the energy the rabbits represent. And predation can have additional, indirect effects. If a predator depends on one species of prey, the prey may survive indefinitely, with only limited population fluctuations, but if the predator uses two prey species, one may be reduced to extinction. In this case, one species of prey becomes extinct because of the presence of the other, even though the two may never meet. Is this competition? Disease and parasitism may have similar effects; species that tolerate them may eliminate or repel species that do not. For example native African ungulates tolerate tsetse-fly-carried trypanosomes, which are fatal to some other ungulates (including cattle brought by man), and which may keep the others out of Africa. This can be considered (indirect) competition. Mayr, quoting Haldane, calls disease a "weapon of competition."

Attempts to answer questions like these, and to define competition precisely, do not seem to clarify the subject. Recognition of different categories of competition seems more likely to do so.

The broadest category, *extended competition,* includes competition in the strict sense extended and modified by all related processes and interactions, including predation, parasitism, disease, and even cooperation; cooperation is a "weapon of competition" if success of the cooperators reduces the success of competing forms. This totality of interactions is Darwin's "struggle for existence."

A useful, evocative term derived from "extended competition" is *competitive repulsion* (compare physicists' "repulsion" among atoms), proposed for the sum of all competitive forces that determine the spacings (including ecologic spacings) and thus the numbers both of individuals in populations and of populations everywhere. Competitive repulsions determine biotic equilibriums, which are considered in Chapter 7.

Extended competition is inconceivably complex. At the other end of the complexity scale is *single-resource competition.* And between these, intermediate in complexity, is *range-limiting* or *niche-limiting competition,* which includes all competitive interactions that directly limit the spatial and ecological ranges (niches) of competing species. And competition can be divided in other ways: as intraspecific or interspecific (for which I prefer the simpler terms *in-species* and *out-species* —compare "inbreeding" and "outbreeding"), and as *density-dependent* or *density-independent.* These terms should be self-explanatory.

In-species competition must occur among the individuals of every species at least occasionally if not in every generation, and it must be in part density-dependent, for only density-dependent competition that increases as individuals become more numerous and decreases as they become fewer can hold populations between upper and lower limits. And out-species competition too is presumably density-dependent—related to numbers of species as well as of individuals—in ways that are difficult to measure but can be estimated.

Measurement and Estimation of Competition. The "basic competition equation" used by mathematicians is

$$\frac{dN}{dt} = rN \ \frac{K - N}{K}$$

For explanation of it, see Wilson and Bossert (1971, page 162), or any similar text.

This equation describes the theoretical effect of competition on the increase or decrease of numbers of individuals in a population. It is therefore sometimes called a population-growth equation. But it does not quantify competition and cannot be used to compare amounts of competition in different species or in different places.

There probably is no practical way to measure amounts of competition in real cases, but naturalists can make generalizations about the distribution of it based on intuition, estimates, and to some extent inferences from visible geographic and evolutionary patterns, as follows.

The Distribution of Competition in Space and Time. In limited situations and for short times organisms may live and evolve without competition. For example, when new areas were uncovered by the melting back of Pleistocene ice in Europe and North America, carabid beetles spread into them, and the first few scattered individuals may have been free of competition. However, as individuals multiplied, in-species competition presumably soon began, and arrival of additional species and species-splitting presumably soon initiated and then increased out-species competition.

This example, in which competition is briefly absent and then relatively simple, but increasing, can be considered as at one end of a geographic-climatic gradient, the other end being in large, stable, favorable tropical areas, where competition presumably reaches maximum complexity and intensity. This gradient will be considered again in relating evolution to area and climate in Chapter 7.

Competitive Extinction and Exclusion. Darwin (*Origin*, page 110) thought that species that "stand in closest competition" will affect each others' numbers, and that some will become extinct. This idea has been formalized by mathematical ecologists into the concepts of *competitive extinction* (Gause's principle, that two populations cannot coexist if they compete for a vital resource limitation of which is the direct and only factor limiting them both) and *competitive exclusion* (by which one species excludes a competitor from a niche). These concepts are correct in simple theory, but the generalizations are not so obvious in the complex world that naturalists see, and cases that seem to fit mathematical theory may not be what they seem. For example, when ecologists find a variable resource divided among coexisting competitors, they assume that the division is necessary for coexistence. This is implicit in May and MacArthur's model of division of a "one-dimensional resource spectrum" (Chapter 3). But the division may really be a secondary effect of coexistence and of the competition and competitive repulsion that result, which may be expected to push competing species apart along "resource spectrums" even if the resource is sufficient for them all. An analogy should make this point. Imagine two apple trees, each well loaded with fruit, and two apple pickers. If the pickers cooperate, they may pick from the same tree, and both may gain by doing so. But if they compete, they will probably gain by picking from different trees, even if each tree has more than enough fruit for both. This is competitive repulsion, and it results in division of the resource, but the division is a convenience rather than a necessity.

De facto Coexistence of Competitors.   Coexistence of species competing for a vital resource is conspicuous everywhere around us. The resource is light. Sunlight is a vital resource for green plants, and competition for it is intense. Nevertheless, large numbers of green-plant species coexist. How do they do it?

The leaves of most green plants absorb the same wavelengths of light. When, therefore, 375 species of trees reach the canopy in 23 hectares (less than $\frac{1}{10}$ square mile—about $0.25/km^2$), of tropical rain forest (Richards, 1969, pages 149–153), they cannot be supposed to divide the light into hundreds of different fractions. The plants, in fact, divide not the light but *themselves and their environment* into fractions by complex processes of coevolution. The fractions of the environments are commonly called "niches."

Niches.   The niche concept is useful for many purposes, but it can be dangerously misunderstood. Even some biologists write as if they think that species in adjacent niches are isolated from each other and do not compete or "avoid competition." In fact, of course, these species do not escape competition but yield to it and use it; competition has put them in their niches, continuance of competition keeps them there, and competition is the means by which niche occupiers keep other species out of niches.

What must be understood about niches is that they are not previously existing pigeonholes with boundaries (the concept of niche boundaries is mathematical rather than biological), but are made and continually modified by the organisms that occupy them; their limits are largely determined by competition; and the whole complex ecologic structure that results is flexible and capable of evolving both in detail and as a whole.

Niches as Ecologic Ranges. The nature of niches can be further clarified by considering them as ecologic ranges (I hesitate to call them *ecoranges*) and by comparing them to geographic ranges. A given piece of land, say a continent, has geographic dimensions and is potentially able to support a set of species, each of which will have a geographic range. The ranges are areas occupied by populations. They do not exist until the populations occupy them, and they continually change as the populations change. Attempts have been made to divide continents into idealized (and therefore typological) geographic ranges corresponding to climatic zones or major biomes, but (as a zoogeographer) I know that the ranges of actual species often do not fit the idealized patterns very well. Similarly, a given piece of land has ecologic dimensions and is potentially able to support a set of species, each of which will have an ecologic range. These ranges are ecologic areas occupied by populations. The ecologic ranges do not exist until the populations occupy them, and they continually change as the populations change. To think of them as "niches" invites (typological) idealization and invites also such confusing concepts as "empty niches," "changing niches," "overlapping niches," and "discontinuous niches."

More broadly, a species' total range may be considered to include interdependent geographic and ecologic components both of which are products of the organism's history, characteristics, behaviors, and interactions with other organisms. The whole is unstable and continually changing.

The Never-Ending Role of Competition. Competition must be a never-ending force propelling evolution even in stable environments. Evolution is a step-by-step-by-step process, each step depending on the preceding ones and making new ones possible, and the evolution of one group of organisms must often force continuing evolution of competing groups (the "Red Queen effect," Van Valen, 1973, pages 17ff with note 32). The evolutionary progressions from, for example, amphibians to reptiles to mammals, or from gymnosperms to angiosperms, were evidently propelled primarily by forces, including competition, within and among evolving populations, not primarily by changes in the environment. Some evolutionists, especially botanists and some paleontologists, overestimate the importance of geographic or climatic revolution in the origins of major new groups of plants and animals. The environment does play a directing and limiting role, but the force of evolution comes from the living organisms, generated by multiplication, *competition*, and selection. Climatic revolutions and continental drift sometimes offer new special opportunities, but plants and animals evolving in advantageous new directions need not and do not wait for them.

Selection

Selection occurs among organized sets at all inorganic and organic levels (Chapter 4), but is most conspicuous and familiar at the Darwinian level. The general principle, at this level as at others, is that selection makes nothing but can operate only after sets have been made by action of their members, or by other processes; and natural selection (in contrast to selection by man) can make no positive choices of what is advantageous but can only eliminate what is disadvantageous.

Biologists as well as other persons find it difficult not to think and speak of natural selection as if it involved positive choice. We habitually say, for example, that in cold climates selection is "for" a warm coat of fur in mammals. This is a carry-over from everyday usage. When we buy a winter coat for ourselves, we directly choose it for its warmth, and it is not easy to remember and to make clear in appropriate words that natural selection cannot directly order a warm coat for a rabbit, but must wait until a rabbit with a warm coat happens to be made by random genetic events, and then can act only be eliminating rabbits with inferior coats. And evolution of the rabbit's warm coat may require successive genetic events (mutations), each followed by costly eliminations.

Although the principle is simple, the concept of selection, like that of competition, is handled imprecisely, inconsistently, and sometimes unrealistically by many evolutionists. Biologists' definitions of natural selection are in fact diverse. Darwin (*Origin,* page 81) says, "This preservation of favourable variations and the rejection of injurious variations, I call Natural Selection." Dobzhansky (1970, page 2) says, "The essence of natural selection is differential reproduction . . . " And Wright (1969, page 28) says, "Selection is here defined as any process in a pop-

ulation that alters gene frequency in a directed fashion [without mutation or immigration]." Each of these sample definitions is composite,
and together they include four interrelated but distinguishable concepts: (1) differential elimination of individuals; (2) differential survival
or preservation of individuals, which may result from or accompany differential elimination, but is not the same thing; (3) differential reproduction, which is stressed by many modern biologists; and (4) alteration
of gene frequencies, which is the definition preferred by mathematical
evolutionists.

By defining selection as differential reproduction or as processes that
alter gene frequencies, biologists not only lump several subprocesses
but also arbitrarily restrict the concept of selection to living levels and
prevent further conceptual analysis and comparison between levels. A
definition that is more precise and compatible with selection at other
levels is required if the components of Darwinian evolution are to be
distinguished, and if Darwinian evolution is to be compared effectively
with processes at other levels. For these purposes, Darwinian *selection*
is best defined as *differential elimination* of individuals and lineages of
individuals. This definition is simple, consistent with the multi-level
principle just stated, and a true description of the process.

At the Darwinian level, selection occurs *both* by differential elimination of individuals *and* by differential reproduction, which results in differential elimination of lineages. To stress differential reproduction
alone is a distortion. It is true that individuals that survive have no place
in evolution unless they reproduce, but it is also true that individuals
that reproduce have no place in evolution unless their offspring survive.
Survival and reproduction are the two legs of the evolutionary ladder.
Both are essential. If there is a quantitive difference, selection of individuals—the premature deaths of uncountable billions upon billions of
individuals—has surely been more directly effective in determining the
details of structure, function, and behavior of plants and animals than
differential reproduction has.

A reasonable speculation is that, by defining selection as differential
reproduction or alteration of gene frequencies, population biologists
and geneticists have inadvertently concealed the essential role of *elimination* in evolution by selection, and have therefore failed to appreciate
the cost and the probable imprecision of adaptation. If so, the biased
definitions of selection are the prime cause of a serious gap in most
modern presentations of the evolution theory.

The Force of Selection and the Energy of Evolution.   "Selection" in ordinary usage implies positive choosing, which may seem to generate

force, although the force actually originates in the mind that does the choosing and in the action of the chooser. But natural selection, which acts only by negative elimination, cannot generate force. Nevertheless, selection does exert force. The phrases "force of selection" and "selection pressure" are common in evolutionary literature, and the direction and rate of evolution surely are determined, in part, by the direction and intensity of selection. This paradox is resolved by an analogy. An hydraulic ram is a device that lifts a small volume of water by diverting part of a larger column of moving water. The ram generates no force of its own; it merely draws on and directs energy derived from a larger source. Selection is the ram of Darwinian evolution. Without generating any force of its own, it draws on and directs energy from a larger source in such a way as to determine, in part and within limits, the direction and rate of evolutionary change. The larger source of energy, which Darwinian selection directs, is provided by continual multiplication of individuals by replication and rereplication of DNA, which is a molecular process. The force of selection at the Darwinian level therefore is derived from molecular, ultimately chemical processes at lower levels.

Categories of Selection.   Darwinian selection, broadly defined as differential elimination of individuals and lineages, can be divided.

Selection acts on both environmental and genetic variation, but only the latter is directly concerned in ongoing evolution.

Selection may be either *competition-independent* or *competition-dependent*. It is the former when cold eliminates cold-susceptible individuals from a population of noncompetitors. However, competition-dependent selection is more important in Darwinian evolution.

Competition-dependent selection may be either *in-species* or *out-species*. In the latter, although it is individuals that compete and are differentially eliminated, groups and species may be differentially eliminated too, by death or failure to reproduce of all the individuals in them.

Stabilizing, Directional, Diversifying, and Disruptive Selection.   In-species selection may be either stabilizing or directional. *Stabilizing selection* eliminates divergent individuals from a population, and prevents change. *Directional selection,* of course, results in directional change. In most populations selection is presumably partly stabilizing and partly directional, resulting in change only in favorable directions.

Diversifying selection may occur when selection favors change in more than one direction in one population. Allele equilibriums (Chapter 5) permit it, or at least do not offer much resistance to it. And diversity-ing selection becomes *disruptive selection* when it divides a population into well-differentiated classes of individuals. This may happen without destroying a population's coherence; the evolution of two conspicuously different sexes in many Mendelian populations is an example. Or, in theory at least, disruptive selection may end by dividing a population into genetically separate parts which occur together as "sympatric" species (Chapter 6C).

Density-Dependent Selection. Selection may be either *density-independent* or *density-dependent*. It is the former when lethal mutations eliminate individuals regardless of the presence or absence of other individuals. It is the latter (density-dependent) when the intensity of selection changes as numbers or spacing of individuals change. This kind of selection is related to, and usually a result of, density-dependent competition.

r and K Selection. Selection in which quick and copious reproduction is decisive is *r selection* (*r* being the symbol for rate of reproduction or of population increase); selection in which individuals' ability to compete for resources and stay alive is decisive, *K selection* (*K* being the number of individuals the environment can carry). As a rule, *r* selection is intense in new situations where there is an advantage in getting there first with the most; *K* selection, in old communities where individual plants and animals defend complex prepared positions (prepared by coevolutions, of course). Garden weeds are *r*-specialists; tropical rainforest trees, *K*-specialists. Both kinds of selection presumably occur in every natural population, but their relation—*K/r*—varies and evolves. *K/r* is likely to be low in an evolving lineage that invades a new "open" situation, and to increase as the lineage becomes part of an increasingly complex evolving association, so that

$$(K/r)_1 < (K/r)_2 < (K/r)_3$$

Fitness and Selection. Fitness is related to selection; the familiar phrase "survival of the fittest" means that, during selection, the fittest have the best chance of surviving while the less fit are eliminated. (But this phrase is often misunderstood; see the discussion that follows and Chapter 2).

Unfortunately, "fitness" is now used in two ways: in one way by mathematicians and in another by naturalists.

Mathematicians designate fitness by the letter *W* and give the (usually hypothetical) genotype (or individual) that, if it existed, would have the best possible combination of the genes actually present in the population an arbitrary fitness *(W)* of 1—which does not mean that it would be sure to survive, but only that it would have the best chance of surviving during selection in the population's environment. Then, all other genotypes (or individuals) are given values of 1 − *W*. This makes fitness *(W)* proportional to rates of elimination and survival of different genotypes, and allows calculation of rates of gene substitutions. See again Wilson and Bossert (1971), or any similar text, for more-detailed explanation.

This usage is limited in ways that are not immediately apparent. It makes comparisons difficult; all organisms are arbitrarily assigned the same base fitness—*W* = 1—regardless of actual selective losses, regardless of whether or not the organisms are well adapted (fitted) to their environments, and regardless of whether or not they are likely to survive. For example, a clone of genetically identical individuals has an absolute fitness (*W*) of 1—all individuals have the best possible combination of genes existing in the population—but may be poorly adapted to its environment and on the verge of extinction. Comparison of the fitnesses (*W*) of different organisms shows only which of them is further from its best possible combination of existing genes, and which for this reason is under more intense selection; it does not show what the organisms' present or future prospects are. This is equivalent (by close analogy) to saying that Jack is further than Mary from realizing his potential abilities and is improving more rapidly, but without saying what either the present or potential abilities of either Jack or Mary are, or whether either person is fitted to fill a particular position.

The mathematical treatment of fitness *(W)* has the further, unrealistic consequence that appearance of a new advantageous gene in one individual in a population *decreases* the population's fitness by reducing the fitnesses (1 − *W*) of all other individuals in the population. As a naturalist I find it more realistic to think of the addition of an advantageous gene as increasing a population's fitness by increasing its chances of surviving.

The other way of using "fitness"—the way Darwin used it and the way we use it in everyday language—is as an attribute of individuals or species that are well adapted to their environments. In multi-level terms, it is proportional to chances of survival of any set at any level. This kind of fitness may be designated *F*. Fitness (*F*) cannot easily be treated mathematically. It varies among sets at any level, and it changes as either a set, its environment, or both change. Since the environment of a set is usually composed partly of the higher set of which it is a member, the fitnesses (*F*) of sets at different levels are interrelated: the fitness *(F)* of a gene in a genotype, of a genotype (or individual) in a gene pool, and of a gene pool (or population) in a set of gene pools (or set of species) are interrelated, so that

$$F_{gene} \leftrightharpoons F_{ind} \leftrightharpoons F_{pop} \leftrightharpoons F_{set\ of\ pops}$$

Sexual Selection. Evolution by selection of two separate sexes is treated in a general way in Chapter 5, under "Bisexual reproduction." Now to be noted are some special aspects of it.

Darwin (*Origin,* page 88) defined *sexual selection* as depending " . . . not on a struggle for existence, but on a struggle between the males for possession of the females [of course it can be vice versa]." In some cases females seem to make positive choices among competing males, but the result is differential elimination of the genes of unsuccessful males. This kind of selection is involved in evolution of characteristics—of form, color, sound, odor, or behavior—confined to or emphasized in one sex or the other, usually most conspicuously in the male. Naturalists see examples everywhere, especially in birds and insects.

Sexual selection in Darwin's narrow sense is only part of what Darwin himself (*Origin,* page 87) referred to as processes by which " . . . natural selection will be able to modify one sex in its functional relations to the other sex" without choices being made; examples (mine, not Darwin's) are evolution of pollination mechanisms in plants (which cannot make choices) or of motile sperm in animals.

Sexual selection may have adverse secondary effects. For example, if bright color in male birds evolves by selection because it increases their chances of mating, it may also make them susceptible to counterselection by increased predation. Or if the enormous antlers of the extinct Irish Elk (which Americans would call a giant moose) evolved by sexual selection (which may have been the case even if antler size was allometric, if the large antlers contributed to sexual success), they may also have interfered with the animal's nonsexual activities and contributed to its extinction.

Sexual selection, especially in mating behaviors, may establish reproductive barriers within and between populations and may result in species-splitting (Chapter 6C).

For more detailed treatments of sexual selection, see the books cited under "Bisexual reproduction" in Chapter 5.

Actual Examples of Selection in Nature. Stabilizing selection (as we now call it) has been observed by naturalists since Darwin's time but no examples of directional selection in wholly natural situations were known to Darwin. Some are known now, but most of them are inferred rather than directly observed, or occur in experimentally manipulated populations. The inferences and experiments are important, in fact conclusive, but naturalists would like to be able to see directional selection occurring in entirely natural situations untouched by man, and the cases thus far described are very few.

Probably closest to the naturalists' ideal of visible selection in populations not manipulated by man are cases of "industrial melanism." In these cases smoke from industrial cities has darkened tree trunks by killing pale lichens as well as by depositing grime, and formerly pale-colored moths that spend the day resting on the tree trunks have evolved melanic forms by selection. Pale individuals that were hard to see on pale tree trunks became conspicuous on darkened trunks, while dark individuals became inconspicuous, and birds found and eliminated more of the pale ones. This was seen happening; a moving picture was made of the birds doing it. In some cases smoke abatement has now allowed tree trunks to be pale again, and the direction of selection has been reversed; moth populations are becoming pale again. For further details and photographs see Kettlewell (1973).

Carabid beetles provide examples of selection-in-progress in which both environments and populations are natural, not significantly changed by man. Carabids are described at the end of Chapter 3; the pertinent fact now is that populations of some species are dimorphic, including both winged flying and vestigially winged flightless individuals. This invites strong selection, which is sometimes visible.

Dimorphic carabids show consistent patterns of distribution in parts of Europe and North America that were glaciated late in the Pleistocene. Populations of many of the species apparently survived the last ice-advance in unglaciated places within or just outside the ice sheets, and selection evidently then favored flightless individuals; flying ones presumably fell and died on the ice. But when the ice melted, many species spread into the newly uncovered areas, and selection during spreading strongly favored winged, flying individuals, which Lindroth calls "parachutists." Figure 3 illustrates the kind of pattern that this produced, and that is evidence of strong, reversed selection. That the boundaries between the areas occupied by short- and long-winged individuals are not abrupt probably means that some "parachutists" mated with short-winged individuals before dispersing.

Selection becomes more directly visible among carabids in special situations. For example, winged carabids including winged individuals of dimorphic species can be seen escaping floods by flying, leaving flightless individuals to be selectively eliminated by drowning or predation—ants and other predators soon clear them out of flood debris. Simi-

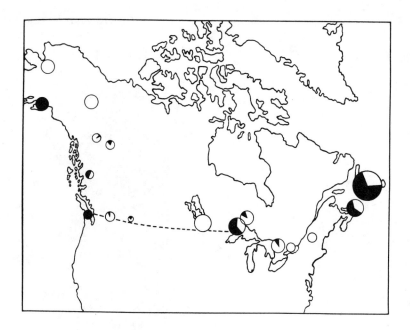

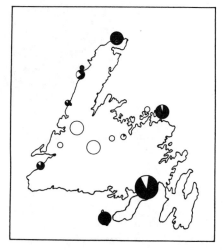

Figure 3. Distribution of wing forms of *Bembidion transparens* Gebler across North America and on Newfoundland (redrawn and simplified from C. H. Lindroth, 1963, The Fauna History of Newfoundland Illustrated by Carabid Beetles, *Opuscula Entomol.*, Suppl. xxiii, pp. 94 and 91). Areas of circles are proportional to numbers of individuals collected at each locality or set of adjacent localities; black parts of circles represent short-winged, white parts long-winged individuals. The geographic patterns suggest survival chiefly of short-winged populations coastwise and perhaps west of the Great Lakes during the last Pleistocene glaciation, and strong selection for long wings and flight during dispersal afterward.

101

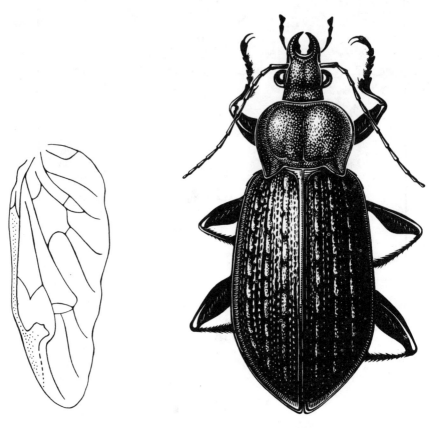

Figure 4. *Carabus serratus* Say: a short-winged individual (wing vestiges not shown) and long wing of another individual to same scale. Actual length of insect, 17 mm.

lar selection can be seen happening on open arctic-alpine mountain tops that are isolated by barriers of heavy forest, for example on the open summit of the Presidential Range (Mt. Washington) in New Hampshire (Darlington, 1943). On warm days in early summer the summit is bombarded by thousands of winged insects carried up by winds from lower open country. Some dimorphic species of Carabidae are included, but of course only winged individuals, this being an example of very strong selection during invasion of an isolated habitat. Usually these species do not establish themselves in the arctic-alpine area, where the climate is severe and the resident carabid fauna, although small, perhaps preemptive, but they may become established on other mountains, or even briefly on Mt. Washington, as an example shows.

The example is *Carabus serratus* Say, which is widely distributed in more or less open country at low altitudes in temperate North America. It is dimorphic; most individuals are short-winged, about 4% long-winged in the lowland population. When I compiled a list of the Carabidae of the Presidential Range (1943), only one individual of this species had been found there, taken by myself under a stone well above timberline, and predictably it was long-winged. However, in 1942 I found several individuals, all within a few hundred feet of each other, a little above timberline on a spur of the range (Boott Spur), and all were short-winged. This was just before I entered the army. I returned to the same place after the war, several years later, but could find no more individuals. Presumably a winged female (homozygous for the recessive allele that allows development of wings) reached Boott Spur, but had mated with a short-winged male (carrying the wing-shortening dominant allele) before flying onto the mountain. The offspring were therefore all short-winged heterozygous individuals, which completed their development but failed to establish a permanent population on the mountain, either because the mountain climate was too severe or because of the presence of a preempting competitor, the arctic-alpine relict *Carabus chamissonis* Fischer.

I do not want to imply that these cases are either fully observed or fully proved. But cases like these do occur in which powerful selective forces are acting now. They are potentially observable, and, although we still depend largely on inference, we can begin to see at least the edges of selection-in-action where neither the populations of the insects nor the environments have been significantly modified by man.

Supplementary Processes

The main process of directional evolution by natural selection is supplemented by processes which are not themselves directional, or not selective, or not primarily at the Darwinian level.

Genetic Drift. *Genetic drift* is change in gene (allele) frequencies not by selection but by chance, as the genes are "shuffled" during reduction division and recombination in sexually reproducing populations. The result may be considerable random change in gene frequencies, and if the populations are small, some genes may be entirely lost.

Genetic drift may have two theoretical results. Small populations are likely to suffer cumulative losses of genes which reduce their fitness; this may be one reason why small populations on islands so often become extinct. And random changes of gene frequencies may happen to carry a small population into a new adaptive zone and initiate a new segment of adaptive evolution. However, this possibility has perhaps been overestimated. Genetic drift results mainly in gene losses, and losses are not likely to preadapt populations to new situations.

Random deviations from Mendelian frequencies may help to establish new advantageous mutant genes in large populations before selection begins. An example is the (hypothetical) establishment of a gene for responsive altruism (Chapter 7). This is genetic drift in a sense, but to lump it with cumulative loss of genes in small populations would be confusing.

The Founder Effect. The *founder effect* can be thought of as a special case of genetic drift. When one or a few individuals leave a large population and found a new one in a new place, the founder(s) will probably carry a very incomplete set of the genes of the large population, and the genes not carried will be lost. The result will probably be that the new population will differ from the old one in many details. For example, if the fertilized female *Carabus serratus* that reached Boott Spur had founded a population, the latter would have lacked all the species' genes that happened not to be carried by the founder or the male she had mated with.

Additional processes that supplement evolution at the Darwinian level occur both below and above it: below, recurrent mutation and molecular evolution (Chapter 5), and above, group selection (Chapter 7).

SECTION B. THE PROCESS AS A WHOLE

This section is concerned with characteristics of Darwinian evolution as a whole (rather than with its components) and with its inherent patterns, rates and modes (whether by gradual change or sudden events), and costs. The rates and modes of evolution are under intense scrutiny now and are examined critically here because naturalists are in a special position to see and perhaps contribute to the understanding of them. And the costs of evolution are not under such intense scrutiny but ought to be; failure to appreciate the heavy cost of evolution by selection—by selective elimination—and the resulting probable imperfection of adaptations may be the most serious error in the "synthetic theory" of evolution.

Characteristics of the Process

Programming of Darwinian Evolution. The direction of evolution at the Darwinian level is mainly determined by natural selection, but details are programmed in additional, less obvious ways. Darwinian evolution is limited as well as energized by events at molecular levels, notably by the kinds and numbers of mutations that occur, which must affect both the rates and directions of evolution, although the effects are difficult to detect and quantify. The environment too limits what evolution can do; land animals could not have evolved until the environment included land plants, nor predators before it included prey. And evolution must be a step-by-step process in which each step limits and to some extent programs the following one. Evolution could not have made a protozoan directly into a fish, nor a fish directly into a man, but could perform these miracles only by successive stages at each of which available materials were used in new ways, and at each of which new materials (pre-adaptations; Chapter 6C) were made available for the next stage.

All these factors together do program each evolutionary continuity at the Darwinian level. But at the same time the continuities are more or less open-ended. There is no obvious or necessary end to them. At this level, as at others, every lineage must have a *P/OE,* or ratio of programming to open-endedness, which cannot be quantified but can sometimes be compared with other *P/OE's*, as noted in Chapter 4.

Evolutionary Choices. Programming limits evolutionary choices but does not forbid them. A "choice" is made, unconsciously, when an evolving organism has two or more alternatives and "chooses" one of them. One such choice was perhaps made during the origin of life, when (and if) right-handed rather than left-handed DNA was chosen for the making of living molecules, and innumerable additional choices have been made since then. Some have been made at random, so far as we know, while others have been determined by immediate advantage. Often when an evolving group has divided, different derived groups have made different evolutionary choices, one of which has turned out to be pre-adaptive and the others not. This has happened many times during the evolution of man's ancestors (Chapter 9). Most biologists see no foresight or purpose in this, and none is implied here; "choice" is used only for want of a better word with the required meaning. An evolutionary choice is like the choice we make when someone offers us a fan of cards back-to, and says "choose one," and we do it blindly, without knowing what is on the card's face. It is even more like the choice we would make if the cards had different symbols on their backs not, or not directly, related to the symbols on their faces, and if we chose a card because the back offered an immediate advantage, without knowing what the longer-term advantages or disadvantages on its face might be.

Whole Individuals and Whole Environments.   Darwinian selection acts not on single genes or single characteristics of organisms, but on whole individuals including their whole repertories of behaviors. And whether a particular kind of plant or animal survives and how it evolves depends not on single factors in the environment but on the whole environment. All evolutionists recognize this principle, but they do not always apply it strictly. Its strict application clarifies some otherwise obscure evolutionary processes. For example, "group selection" is clarified by looking at it as selection of individuals in whole environments that include other members of a group (Chapter 7).

Structure, Function, and Behavior.   Whole individuals are integrated packages of structures, functions, and behaviors; these things are interdependent, and they often evolve together. A process or behavior cannot evolve unless an appropriate structure exists or evolves to perform it. And, conversely, a special structure does not, we think, evolve unless a selectively advantageous use evolves for it. When, therefore, the fossil record shows that a structure or organ has evolved conspicuously, we suppose that it was useful, and we try to decide what the use was. We do this when we try to understand and explain the rapid evolution of the human brain (Chapter 9).

Irreversibility.   Evolution is never reversed precisely; this is "Dollo's law," based originally on the fossil record, but inherent in the concept of step-by-step evolution. A single simple genetic change may sometimes be reversed by a back mutation or reversal of selection, but precise backtracking along a complex series of steps is impossibly unlikely. When, for example, a terrestrial mammal, derived from a fish by a long series of evolutionary steps, has returned to a fishlike existence—that is, become a whale—evolution has not backtracked but has followed a new step-by-step continuity to a product which is superficially like a fish but very different in general organization and innumerable details.

But note: "Dollo's law" does *not* say that a specialized organism cannot lose its specialization and become ancestral to a differently specialized group. Loss is not the same as reversal. The (hypothetical) evolution of a specialized tunicate into a primitive chordate after loss of adult specializations (Chapter 8) is an example. The frequently used argument that *A* cannot be ancestral to *B* because *A* is specialized is fallacious.

Rates of Darwinian Evolution

Rates of evolution can be measured by rates of gene (allele) substitutions, rates of speciation, or simply estimated amounts of visible change: the last is the most difficult to quantify but is also the most widely applicable and easiest to appreciate. And the time scales against which rates of evolution can be measured are the geologic timetable (see Chapter 8, Figure 6); events that can be dated, including the beginning and ending of glaciations and the formation of islands; and in special cases the amount of change of supposedly nonadaptive proteins.

Horseshoe Crabs and Opossums.   Rates of evolution are extremely variable.

Existing horseshoe crabs belong to a genus, *Limulus,* which appeared in the fossil record in the Jurassic well over a hundred million years ago, and which has evolved very little since then, in a time span twice as long as the Age of Mammals. And the genus *Limulus* itself is "not so far removed from still more ancient members of this general stock which trace back to the most ancient of arthropod times in the Cambrian . . . " (Romer, 1968, page 74), 500 million years ago.

Among mammals, American opossums, family Didelphidae, are fossil in the late Cretaceous. The genus *Didelphis,* which includes our existing North American opossum, happens not to have been found fossil before the Pleistocene, but is very similar to late Cretaceous forms. Didelphids have existed, evolving relatively little, through the whole of the Age of Mammals, while successions of families and orders of placental mammals have evolved, radiated, and replaced each other in the main part of the world, and while marsupials other than didelphids have radiated conspicuously in Australia.

Insects in Amber.   Insects fossil in Baltic amber provide less familiar examples. The amber—fossilized gum of various trees—was deposited in northern Europe something like 40 million years ago. Some insects, even some *species* of ants (Wilson, 1955), have changed so little since then that they are difficult to distinguish from living species, while other insects have apparently evolved rapidly. For example, Carabidae in the Baltic amber include a number of paussids (tribe Paussini), but all the amber paussids belong to primitive subtribes. Not one of the now-dominant and strikingly diversified existing groups of derivative paussids has been found in the amber, and this suggests an explosive evolution of genera and subtribes as well as of species of derivative paussids during the middle or late Tertiary (see "Paussids" and Figure 7 in Chapter 8). The derivative paussids are obligatory ant guests, and the contrast between the slow evolution of at least some ants and the apparently explosive evolution of their paussid guests is extraordinary.

Interglacial and Hawaiian Carabidae. Carabids offer other significant comparisons. Fossil fragments of many species of carabids have been found in interglacial peat, deposited during warm intervals between Pleistocene ice-advances in Europe and North America. Many of the fossils were once thought to be different from existing species, and rapid late-Pleistocene evolution was postulated. However, it is now thought (because of improvements in the taxonomy of existing species and discovery of microscopic recognition characters that are visible in the fossils too) that there has been no detectable change—no visible evolution at all—in any of the Carabidae in question since the last interglacial, something like 100,000 years ago (Coope, 1978).*

In contrast is the situation on the Hawaiian Islands, where carabids belonging to some of the same groups that are well represented in the interglacial peat in Europe and North America (for example, *Bembidion* and Agonini) have diversified explosively, evolving scores or hundreds of new species and many new genera. The time involved is unknown, but the difference between the nonevolution of the interglacial carabids and the explosive diversification of some of the same groups on the Hawaiian Islands is remarkable and probably reflects great differences in evolutionary pressures and rates.

Hawaiian Moths. The Hawaiian Islands are remarkable for other examples of striking and probably rapid diversification of birds, land snails, and insects, but in most cases they cannot be precisely dated. However, one small group of Hawaiian moths has apparently speciated within the last 1000 years: 5 related endemic species within the pyraustid genus *Hedylepta* feed only on the banana, which Polynesians introduced into the Hawaiian Islands only about 1000 years ago (Zimmerman, 1960).

* In a letter dated May 29, 1979, Professor Coope says, ". . . all known fossils [of Carabidae] from [the Pleistocene] can indeed be assigned to existing species. Work on Late-Tertiary [Canadian] insect faunas by J. V. Matthews . . . has shown that there is only the slightest of difference between . . . specimens several million years old and their modern equivalents . . . [while] during this same period the fossil record of the mammals shows so much evidence of evolution."

Explosive Evolution. Explosive evolution, as evolutionists including Simpson (1944, pages 89, 139, 213) usually describe it, occurs when, from a common ancestor, many different lineages evolve rapidly in different directions, becoming adapted to different environments or different ways of living. As thus described, it is equivalent to rapid adaptive radiation. However, explosive evolution may be more than this: it may be a feedback process comparable to a hearth fire (Chapter 4) which is self-propagating, self-accelerating, and self-terminating.

Even relatively simple segments of adaptive evolution may have these characteristics. For example, the evolution of a Batesian mimic might begin if the potential mimic at first resembled the model only distantly, and if the distant resemblance was advantageous only to an occasional individual under unusual circumstances, perhaps only in dim light when an individual mimefactor (predator inducing mimicry) could not see clearly. Such a resemblance would have only a slight selective advantage at first, and its rate of evolution would be slow. However, as the resemblance became closer, its selective advantage would presumably increase, and its rate of evolution would increase. Its evolution would then be both self-propagating and self-accelerating. But as the resemblance became very close, both the possibility and the advantage of further improvement would decrease, and so would the force of selection and the rate of evolution. The evolution of the mimic would therefore be expected first to accelerate and then to decelerate.

The evolution of placental mammals may exemplify a more complex explosive pattern. The primary general adaptation of primitive mammals was probably warm-bloodedness or, more accurately, a shift from behavioral toward increasingly physiological thermoregulation. (This may have permitted and been correlated with evolution of placental reproduction, a better brain, and other general adaptations.) The detailed evolutionary changes in this direction were perhaps slight at first and conferred only slight selective advantages, so that they evolved slowly. Division of the palate in some Triassic fossil therapsid reptiles suggests that the ancestors of mammals were beginning to become warm-blooded then (Romer, 1968, page 237). If so, the earlier stages of evolution of mammalian warm-bloodedness may have proceeded slowly for something like a hundred million years, from the Triassic to the Late Cretaceous; the complexities of the homeostatic temperature controls may well have required a long time span for their evolution. Then, this segment of evolution apparently accelerated, or at least reached a point at which it conferred great advantages. And then, when the complex physiological processes concerned approached effective limits, the force of selection decreased and the rate of evolution of the primary adaptation presumably decelerated. However, the general selective advantage conferred at the climax of the primary adaptive segment of evolution was so great that the animals concerned—Late Cretaceous and early Tertiary placental mammals—radiated explosively in secondary ways.

In more general terms, explosive evolution may involve a feedback cycle of adaptation which begins slowly, accelerates, and then decelerates. But if the primary adaptation is a general one conferring a great selective advantage, it may be followed by an explosive radiation involving secondary adaptations and conspicuous speciations. These two parts of an evolutionary explosion—the primary adaptation and the secondary radiation—should be clearly distinguished. The two are probably complexly interrelated. Evolution of a very advantageous general adaptation is likely to be followed by a very extensive secondary adaptive radiation and multiplication of species; and the diversity of species, each possessing the advantages of the general adaptation plus its own special adaptations and random genetic effects, should increase the chance of a breakthrough into a new adaptive zone and the beginning of a new segment of general adaptation.

Generalizations about Rates of Evolution.   Simpson's old (1944) summary of rates of evolution is still valid: rates have varied enormously in different groups at the same time, and in the same groups at different times, and most of what we now think of as slow-rate lines of evolution have probably gone through periods of rapid evolution in the past and have decelerated.

In general (not from Simpson): the times required for evolution of new species may range from "instant speciation" in one generation (by polyploidy or some other abrupt genetic change) to perhaps tens of millions of years; and of new genera, from possibly one million (or less) to hundreds of millions of years.

All this suggests a generalization (not a "law"!): evolutionary explosions and rates of evolution in general are, if not cyclical, at least segmental, each well-defined segment of adaptive evolution tending to have a history of acceleration and deceleration, sometimes followed by conspicuous secondary adaptive radiation and speciation. The segments are irregular, and they do not occur at regular intervals, but they do tend to form irregular sequences of adaptations and radiations. The complexity of the whole is inconceivable—literally beyond the power of the human brain to conceive of—but the principles are real and well worth attempting to apply to, for example, the evolution of man and of human cultures (Chapter 9).

For a longer account of rates of evolution as a paleontologist sees them, see again Simpson (1944), and for a shorter but more recent discussion of rates, with additional references, see Darlington (1976).

Modes of Evolution

Mode is, or may be, independent of rate of evolution, although related to it. Two questions can be asked about it. Does the main course of evolution run through large, long-lasting populations, or through small populations that frequently replace each other? And does evolution proceed by a continuity of small steps, or by periods of equilibrium punctuated by sudden "revolutions" (Gould and Eldredge, 1977)?

Paleontologists and naturalists tend to answer these questions differently. We both see the same kinds of species, distinguished by visible characters and not by hidden genetic or molecular ones, but we see them differently. Paleontologists have the advantage of being able sometimes to see apparent evolutionary continuities. But naturalists have the advantage of seeing individuals, populations, and species alive and in more detail.

As a naturalist-taxonomist I see, or think I see, two significant patterns among the carabid beetles and other animals with which I am familiar. One is a pattern of seemingly endless small variations of individuals within and between large populations, consistent with gradual separation of large populations into parts (usually geographic subspecies) which evolve as separate species. Small peripheral populations (for example on islands) may differentiate more rapidly, but usually become extinct. The other pattern I think I see is one of dispersal of evolving plants and animals from large to small areas and from more to less favorable climates (see "Area, climate, and evolution" in Chapter 7), as if *effective* evolution occurred in large populations in large areas, not in small populations in peripheral areas.

These visible patterns are consistent with Wright's generally accepted model of effective evolution in large, deme-divided populations. Compare, for example, a large population of $10N$ individuals divided into 10 $1N$ demes, and 10 small independent $1N$ populations. The small populations may evolve more rapidly and in more directions, but the small demes can evolve rapidly too, and they have an overwhelming advantage: they can put together events that occur in different demes. When, as is probably often the case, a major (general) adaptation such as mammalian homeothermy involves putting together an almost endless series of mutations, recombinations, and separate selective events, the advantage of the large deme-divided population must be almost incalculably great. It is, as will be noted again in Chapter 9, comparable to the advantage that a set of brains has over single brains in solving complex problems. Just as 2 heads are better than 1, and 10 much better if they put together what they have learned, so 10 demes that put their gains together are much better than 10 separate small populations, and better than a large population that is not deme-divided.*

As to the second question, some paleontologists see, or think they see, in the fossil record that evolution consists of periods of equilibrium punctuated by "revolutions." But the idea of evolutionary equilibrium is contrary to the "Red Queen Hypothesis," and "revolutionary" discontinuities in the fossil record may have another explanation. Dispersals (geographic movements) and replacements of populations are continually occurring. They are required by the deme-divided model of evolution; naturalists and biogeographers see them everywhere; MacArthur and Wilson (1967) have described some of them formally; the very existence of swifts implies them (see Chapter 6C); but paleontologists, whose evidence is immobilized in rock, cannot easily see them. Perhaps the discontinuities that paleontologists see in the record of evolving organisms may be due more to replacements-by-dispersal than to "revolutions" in situ.

These considerations suggest that the main course of evolution runs through large, but deme-divided, populations by a continuity of small steps; the process varies enormously in rate (as noted earlier in this chapter), but does not involve equilibriums punctuated by revolutions. The evolutions and extinctions of small, peripheral populations are easier to see but (usually) out of the main stream. And sudden genetic changes, including polyploidy, do occur and may divide species but (perhaps) do not contribute much to ongoing evolution.

There is, incidentally, nothing in Darwin's view of evolution to forbid catastrophes. In fact his theory requires them. Natural selection depends on most individuals being selectively eliminated by individual catastrophes. Extinctions—species catastrophes—were emphasized by Darwin. And he accepted extinctions of major groups and faunas on a catastrophic scale. But he treated them as components or consequences of an essentially orderly evolutionary process.

---

* This principle can be applied to Snow's "two worlds." If, as Snow supposed, our scientific and nonscientific intellectual worlds are evolving separately, with little exchange, this may be good rather than bad, provided that occasional exchanges do occur—and Snow probably underestimated them. Our whole intellectual world is in fact deme-divided in much the way that large populations of, say, carabids or English Sparrows are, and this makes for diversification, competition, and effective evolution of ideas. An intellectually uniform population would be intellectually stagnant.

The Cost of Selection

During the "evolution" of a man-made machine, the maker of it can take it apart and eliminate inferior and substitute improved parts without throwing away whole machines. But in organic evolution, each slightest separate improvement made by natural selection in the great number of parts, processes, and genetically determined behaviors of an evolving organism is made at the cost of eliminating—throwing away— all the whole individuals that lack the improvement. An example, the evolution by selection of a warm coat for a rabbit, is given in Chapter 6A. This is "the cost of natural selection," or of adaptive evolution. Calculation of this cost and of its consequences is one of the most important pieces of unfinished business in modern evolution theory.

Haldane, in 1957, calculated that gene (allele) substitutions by selection cost so much that populations must (in effect) choose between evolving at low cost per generation but uncompetitively slowly, or more rapidly but at a cost so high as to risk extinction. These alternatives constitute what is sometimes called "Haldane's dilemma." Geneticists and mathematicians have suggested complex and ingenious solutions to this dilemma, but have rarely faced what may be its real significance: that the high cost of natural selection may in fact limit rates of adaptive evolution, and that complex adaptations in complex environments that are themselves evolving may usually be, not "maximized," but incomplete and imprecise.

The cost of selection can be clarified, for non-mathematical naturalists, by means of hypothetical cases and simple arithmetic.

The first case is so drastically simplified that it is (I hope!) science fiction. Suppose that, in a population of four billion individuals, only one male and one female are homozygous for a recessive allele which allows them alone to survive an atomic holocaust that eliminates all the other individuals. If all individuals survive, but only the two can reproduce, the others being sterilized by radiation, the effect is the same. Then one complete allele substitution will have been made by selection in one generation, at a cost of the loss of 3,999,999,998 individuals, and no additional *selective* substitutions can be made and paid for at the same time, although other, random losses of genes comparable to those expected of the "founder effect" (Chapter 6A) will probably occur.

And now suppose that the two survivors of the holocaust escape other hazards and produce ten offspring of which two, a male and a female, are homozygous for another recessive allele which allows them to survive, say, a protein shortage that results in the deaths (or failure to reproduce) of the other eight individuals. Then another complete allele substitution will have been made in one generation at a cost of the loss of eight individuals, and again no additional *selective* substitutions can be made and paid for at the same time.

These cases yield two initial generalizations. In simple cases like these, the minimum cost of a substitution of a new for an old allele is the selective elimination of one whole generation of the population, less the (usually few) new-allele bearers; this is true regardless of population size. And therefore a population can pay for not more than one such simple substitution, or the equivalent, per generation.

This cost may be either increased or decreased in more-complex cases. If an allele substitution is paid for in increments spread over several or many generations, the cost may be multiplied many times; this might be called the Haldane effect, since he was especially concerned with it. But if favorable alleles are linked into sets which are inherited as units, costs are reduced; the cost of one substitution remains the same, but one "payment" covers all the alleles in a linked set; it is, by analogy, as if we paid a messenger to carry a package of gifts to our children, the messenger being paid the same whether the package contains one or many items. However, there may be a catch here. Although a linked set of alleles may be substituted cheaply, making up the set may require formation of diverse sets followed by selective elimination of great numbers of individuals with inferior sets, and this process may take more time and cost more than substitutions of the alleles separately. To continue the package-of-gifts analogy, although it is cheaper to send the package as a unit than to send the gifts separately, the cost of preparing the package may cancel out the saving.

This is just the beginning of the complexities of the subject, some of which are considered in a recent paper of mine (1977) that gives additional references. Briefly, the conclusions are that selection should be most cost-effective—that is, should result in most change at least cost—when substitutions of recessive, large or linked, serial alleles occur rapidly and therefore at low cost in small demes in a large population, and are followed by deme-group substitutions. (This is the deme-divided model of population evolution already discussed in this section.) If the individuals in each deme are "linked" by mating preferences or reciprocal altruistic behaviors, and if the demes are not too numerous, the deme substitutions may be rapid and low cost too.

This is mostly theory. How does it fit real cases, and especially how does it fit our own history? We can derive a first approximation to an answer from three propositions: (1) one allele substitution, or its equivalent, can be made and paid for in one generation; (2) the evolution of *Homo* from *Australopithecus*, or from a comparable ancestor, has spanned perhaps a quarter of a million generations (20 years per generation for 5 million years—this is a reasonable guess); and (3) only a small fraction of our genes have changed during this time (judging from comparison of our genes with the chimpanzee's, and assuming we have come from a common ancestor). This would seem to allow up to a quarter million allele substitutions during the evolution of *Homo*, which at first thought would seem to be sufficient to make man from a man ape.

But if we look at man now, he does not seem as well made—not as perfect either physically or behaviorally—as he ought to be. What is wrong with the calculation? There are several possibilities. The cost of each substitution, paid in increments, may have been many times one generation of individuals—the Haldane effect. Or, some genes may have evolved, not by a single substitution, but by steps—by substitutions of successive alleles at the same loci—the full cost of one substitution being paid at each step, over and over again. Or, and perhaps more important, different genes—genes at different loci—may have had to evolve complex relationships among themselves, with changes in some genes continually forcing changes in others—the Red Queen effect at the gene level. Man is not a set of separate genes that evolve independently, but an inconceivably complex set of interacting sets of genes complexly adapted to each other, and it may be beyond the power of evolution by selection to fit the complex whole precisely to the complex and changing environment; the cost may be too great.

Cost and the Imprecision of Adaptation. A species' adaptations evolve by selection, which is to say by gene (allele) substitutions by selective eliminations. This is a slow and costly process at best, and it occurs in an environment that is itself evolving, partly by evolution of associated species. A species' "target" of adaptation is therefore continually moving and never reached, and the species must continue its evolution endlessly if it is to keep its place in its community. This has been called "The Red Queen's Hypothesis" (Van Valen, reference given under "The never-ending role of competition"), after Alice's Red Queen, who had to keep running just to stay where she was. The least that should be inferred from this is that adaptations *may* be imperfect, and that we should look at them to see if they are.

Naturalists can see many indications of adaptive imperfections in the real world. Among the carabid beetles I work on, closely related species that seem to be living in the same way in similar environments usually differ in many small, apparently nonadaptive details. I used to take on faith that these details are precise adaptations to present or past environmental differences that I cannot see. Now, I think they may be manifestations of adaptive imprecision. Similar examples of apparently nonadaptive differences within and between species are conspicuous in many groups of plants and animals; taxonomists often prefer to use what they call "nonadaptive" characters in making classifications. A broader indication is that, if adaptations were perfect, adaptive evolution would stop, but it has not stopped.

Man's own adaptations are imperfect. Our backs, hearts, feet, and so on are not yet fully adapted to the two-legged posture our ancestors assumed probably more than five million years ago, and we are still imperfectly adapted psychologically to our new social environment. Here are two specific examples of man's unadaptedness, for naturalists to think about, one involving the senses, the other a behavior.

A keen sense of smell as well as keen vision should be advantageous to man, and should be maintained by selection, as noted in Chapter 3 under "Mathematics." Nevertheless, our sense of smell is not good. This may be because too much attention paid to odors interfered with our eye-oriented focus during an early stage of our evolution (see "Focusing" in Chapter 6C), or more likely because there is not space on man's brain for a complex odor-detection-and-interpretation system as well as for all the other systems that crowd our brains. There is competition for space on the brain, or, indirectly, competition among the genes that determine competing systems on the brain.

The other example concerns ability to swim. An innate, gene-determined swimming ability would surely be advantageous and might have evolved by selection; elimination of nonswimmers by drowning must have been a measurable, or at least estimable, selective force for millions of years, and still is. Nevertheless, we have not evolved a gene-determined ability to swim. Persons who assume that selection must maximize adaptations and behaviors should remember cases like these.

Why do evolutionists so often assume, explicitly or implicitly, that adaptations must be precise or nearly so—the fashionable word now is "maximized"? One reason may be that many adaptations are astonishingly good; but this does not mean that they are perfect. Another reason is that many evolutionists, especially mathematicians, do not see organisms as evolving wholes, but see only separate parts, sometimes only single allele substitutions, which can be treated mathematically; but completion of one allele substitution does not make a perfect adaptation. But the principal reason may be that mathematical evolutionists are so sure their models must fit reality that they do not look at the real real world to see if they do.

That adaptations are imprecise or imperfect has been suggested by Darwin (cited by de Beer) and emphasized by de Beer (1971, pages 9–11), who calls it "a principle of great importance." Considerations I have summarized in the last few paragraphs suggest that most species most of the time are in fact far from perfectly adapted to their environments, but are just a little better than their competitors, for the time being. If so, the current fashion among evolutionists of assuming that selection must have "maximized" particular reproductive strategies or feeding behaviors, and of basing mathematical models on the assumption, is dangerously unrealistic. Selection evidently can produce striking adaptations (for example, man's erect posture) rapidly, by focusing on them, but the primary adaptations may not be as precise as they seem to us to be, and secondary modifications (for example, those secondary to man's erectness) may evolve much more slowly and much less precisely, because their evolution by selection costs so much.

This simply stated conclusion suggests an analogy. Suppose that a man—a human individual—although alive and functioning, is imperfect and could be improved in 1000 ways. Suppose that he consults 1000 specialists, each of whom prescribes a specific treatment that can make one improvement. And suppose that each prescription is expensive and that all together they cost so much that the man can afford only a few of them. Then, even if all the other prescriptions are correct and able to do what is intended of them, they are no use to the man; he cannot afford them. An analogous situation probably exists in many evolutionary situations: a thousand mathematical specialists may prescribe different things that an organism or population might do to improve itself, but the prescriptions are so expensive (in selective eliminations) that the organism can pay for only a few of them. Then, even if all the other mathematical "prescriptions" are correct, they are of no use to the organism; it cannot afford them. This analogy is not evidence, but it emphasizes a possibility that surely should be seriously considered by evolutionists and that may invalidate much—I cannot guess how much—sophisticated evolutionary mathematics.

SECTION C.   CONSEQUENCES AND EXAMPLES

This third section on evolution at the visible, Darwinian level is concerned with consequences of the process, including adaptation, specialization, and evolutionary "focusing"; evolutionary radiation, parallelism, convergence, and mimicry; diversification, and its relation to stability; races, species, and higher categories; and extinction; and the chapter ends with everyday examples of evolution that are so familiar that naturalists do not often think about what they mean or how difficult they are to understand.

Adaptation

Darwinian *adaptation* is a process, and its products are *adaptations*. It is the process of selection that fits particular organisms to particular environments or ways of doing things, and the adaptations it produces are attributes of organisms—structures, processes, and behaviors—that contribute to fitness. It is a direct and necessary consequence of Darwinian evolution by natural selection, and it constitutes the main volume and thrust of evolution at the Darwinian level, although it is limited, supplemented, and modified by secondary processes.

Special adaptations fit organisms to special environments or special activities. *General adaptations* include improvements of structures and processes that allow some organisms to live and reproduce in many environments. They are adaptations to the general environment of the world, and they lead to success over great areas and in many special situations; mammalian homeothermy (warm-bloodedness, or the maintenance of body temperature by internal homeostasis) is an example. The two kinds of adaptation have much in common: general adaptations are in fact usually sets of special adaptations.

Pre-adaptation.   Literally, a pre-adaptation is an attribute of an organism that originates first and becomes useful later. A favorable gene mutation is a pre-adaptation in that it must originate before it can be useful. But "pre-adaptation" has a broader meaning in evolution. An *evolutionary pre-adaptation* is an attribute—a structure, process, or behavior—that is (1) useful (selectively advantageous) when it first evolves, and (2) useful also later, in a new environment or in a new way. Pre-adaptations like this have been decisive in organic evolution, including the evolution of man as described in Chapter 9. Pre-adaptation is evolutionary opportunism; it does *not* involve foresight.

Specialization and Focusing.   The selective forces that act on an organism are exceedingly complex, but within the complex selective matrix evolution in any one case emphasizes—that is, focuses on—a limited number of adaptations. This is *specialization.* Two facts about it are remarkable. One is its narrowness; organisms are often specialized to live only in very narrow niches or to depend on only one other kind of organism. Examples are insects specialized to live on one part of one kind of plant, and plants specialized to be pollinated by one kind of insect. The other remarkable fact is that few different conspicuous specializations, often only one, occur in any single kind of plant or animal. This concentration of evolution on one or a very few adaptations in each case can be called *evolutionary focusing*. It results in organisms putting maximum effort where it counts most. And one of its effects is that it allows individuals to act as wholes rather than as sets of separate parts. Individuals often must act as wholes, quickly and effectively, and their actions must therefore be focused on one thing at a time. For example, there is no obvious reason why an animal cannot have both acute eyesight and acute power of odor discrimination, but the two do not usually go together, and it may be selectively advantageous to focus on one and make the other secondary. (However, other explanations are possible, including competition for space on the brain, which may not have room for two separate, competing control systems.)

This principle can be extended to details of animal behavior. A diversity of stimuli continually reach an individual's brain, but the individual usually focuses on and responds to one stimulus at a time (of course this is an oversimplification), the selective advantage presumably being that this favors prompt, precise, effective action. The mechanism of this focusing is consciousness, or the brain's power of selecting only a very small part of the stimuli it receives as a basis for action. It is an extraordinary power, which I think is not yet well understood, but which probably has a long evolutionary history.

In man, focusing and consciousness have a further extension: they permit and favor following a single train of thought to a conclusion.

Radiation, Reradiation, and Cumulative Sequences.   The phylogenies of successful groups of plants and animals have usually consisted of successive episodes of diversifications, each episode being called a *radiation* or, since diversification usually involves diverse adaptations, an *adaptive radiation*. Examples are given in Chapters 5 (under "Origin and early evolution of life"), 6B (under "Explosive evolution"), 8, and 9. Each radiation is fanlike; the history of a complexly evolving group can be diagrammed as a series of fans, each fan developing from one ray of the preceding fan (Figure 5). This pattern of evolution is itself a consequence of the Darwinian process, and it has several secondary consequences and raises important questions.

A practical consequence of the pattern diagrammed in Figure 5 is that, when only a few fossils of a complexly evolving group are known,

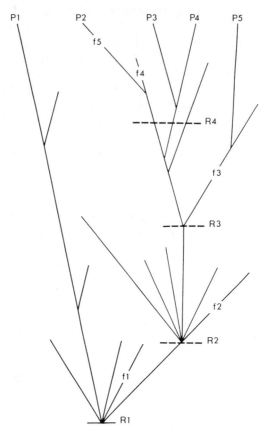

Figure 5. Diagram of an imaginary phylogeny of successive "fans" or radiations (R's), positions of fossils (f's), and existing populations (P's). The radiations are all diversifications after single breakthroughs except R4, in which several rays evolve to a new grade. Of the fossils, only f5 is strictly ancestral to population P2, but the others, although on collateral rays, constitute an informative evolutionary sequence.

they are more likely to represent points on fans (as indicated in Figure 5) than to form a direct linear sequence. This probability should be remembered in relating, for example, the scanty fossil record of primates to the line of evolution leading to man; the fossils are more likely to be man's collateral relatives than actual ancestors. But "collateral" fossils can show sequences and times of evolutionary stages decisively.

Important questions are what actually happens when one ray of an adaptive fan reradiates, and why does it happen? What happens is that the plants or animals representing one ray of the fan evolve an adaptation which allows them to enter a new environment or to do something new. This is a pre-adaptation in the evolutionary sense, which allows a "breakthrough." And if the new environment is diverse enough (for example, the land as a new environment for a fish) or the new activity advantageous enough (for example, flight as a new activity for an insect) a new episode of adaptive diversification follows.

Although the major groups of plants and animals have evolved by successive adaptive radiations and reradiations, representatives of earlier radiations have frequently survived, so that evolution has produced *cumulative sequences* rather than (or as well as) successive replacements.

The pattern of organic evolution, then, has not been simply linear, although we sometimes have to describe it that way. It has been a pattern of successive, complex radiations and reradiations, and of selective extinctions, survivals, and secondary radiations which have produced exceedingly complex cumulative sequences. All this has greatly increased the theoretical and actual complexity of Darwinian evolution and the difficulty of tracing phylogenies.

Parallelism and Convergence. When two or more closely related organisms evolve in the same direction, this is *parallelism*. When distantly related organisms evolve similar adaptations, this is *convergence*. Parallelism occurred when (and if) several closely related mammal-like reptiles evolved similar mammalian characteristics in the Mesozoic; convergence, when unrelated swifts and swallows evolved similar adaptations for catching insects in flight (see later in the present chapter).

Mimicry. Organisms that evolve in parallel or convergently need have nothing to do with each other. They may occur in different parts of the world (the radiation of marsupials in Australia parallels the radiation of placentals on other continents, and this general parallelism includes many special convergences; see Chapter 8), or at different times in geologic history (for example, the radiations of dinosaurs in the Mesozoic and of large mammals in the Tertiary). And parallel or convergent organisms sometimes coexist without affecting each other very much, as swifts and swallows seem to do. However, parallelism and convergence sometimes result in advantageous resemblances, or *mimicry*. Two common kinds of mimicry (there are others) are *Batesian mimicry,* in which edible mimics escape predators by resembling inedible or protected models, and *Mullerian mimicry,* in which two or more protected species resemble each other, so that predators avoid all of them after sampling any one. Mimicry of course evolves by selection (see Chapter 6B, under "Explosive evolution"). To naturalists, it is one of the best evidences of the effectiveness of natural selection, and a fruitful field for selection experiments in nature. The subject is well covered by Wickler (1968).

Note that mimicry can involve not only form and color but also sound, odor, surface texture, and especially behavior. It is often necessary to see mimics alive, and to see how much their postures and movements resemble those of their models, in order to appreciate their mimicry. And the effect of mimicry on predators can be judged not only by observation and experiment but also by our own reactions; even I, an experienced entomologist, have trouble forcing myself to take in hand an insect which I know is a fly (by its single pair of wings and fly's antennae) but which looks and sounds like a bee! Mimics are most numerous in the tropics, but naturalists will find some almost everywhere. The resemblance of the edible Viceroy butterfly to the inedible Monarch is a familiar example, and north-temperate America possesses moths, flies, and beetles that look and sound deceptively like wasps and bees.

Diversification

Diversification is a multi-level process (Chapter 4), but here the emphasis will be on diversification of species and its effect on community stability.

Species Diversification. Species diversification is an inevitable consequence of Darwinian evolution, for two reasons: because the environment is diverse, and because evolution includes stochastic or random components.

Diversification of the environment began before the origin of life as a result of evolution of the earth's surface, which was made up of different elements combined into diverse minerals and chemical compounds, which became differently distributed according to their weight, stability at different temperatures, and solubility in water. Uneven distribution of materials, resulting from complex processes within the earth, divided the surface between land and water. And the shape and motions of the earth made different parts of the surface hot or cold and well-watered or arid, while local factors made a diversity of local environments. Evolving species did and do diversify by adaptation to the diversity of the inorganic environment. And, since species are parts of each other's environments, diversification of species has continually increased the diversity of the environment, which has induced further diversification of species (see again the Red Queen effect, Chapter 6A, under "The neverending role of competition").

The second reason for diversification of species is that when a population is divided into separate "daughter" populations, even if the latter are initially identical and in identical environments (and absolute identity is possible only in theory), they will diversify because chance will introduce different mutations and gene recombinations at different times into the separated populations, and because their environments will change differently too, partly by chance.

Species diversification has theoretical advantages, disadvantages, and limits. The primary selective advantage is presumable increase of competitive efficiency. If a generalist—a species that uses the whole of a variable resource—divides into two or more daughter species specialized to use different parts of the resource more efficiently, each specialist will out-compete the generalist in using a part of the resource, and the specialists will replace the generalist.

Diversification of specialists will tend to result in increase in number and decrease in size of populations, and small size of populations may bring additional advantages—the advantages of being rare. Small populations may be less likely to attract predators or support diseases than large ones are. And small populations may evolve more rapidly and at lower cost than large ones, although probably less effectively (Chapter 6B).

However, both specialization itself and smallness of populations bring risks. Specialists risk extinction if the special resources they depend on become unavailable even temporarily. And small populations risk extinction as a result of local or random events that larger populations could survive. These risks are reduced if the individuals forming small populations are widely scattered rather than densely concentrated; wide scattering of individuals reduces the risks of predation and disease transmission and of extinction by local catastrophes. But wide scattering of individuals also decreases the chance of outcrossing. Plants can partly overcome this difficulty by synchronous flowering of individuals and use of wide-ranging, specialized pollinators. Insects can partly overcome the difficulty by synchronous emergence of adults and evolution of effective chemical (odor) signals. However, in spite of adaptations like these, there must be a minimum size and minimum density below which populations do not long survive, and this limits the diversification that can occur under even the most favorable circumstances.

Although all this has been presented as theory, the theory fits remarkably well the situations that naturalists see in tropical rain forests.

Species Diversity and Community Stability. The relation between diversity of species and stability of communities is being reexamined by ecologists and evolutionists now (see Goodman, 1975). There is an obvious relationship. Species-diverse, complexly organized communities have time to evolve only in environments that are sufficiently stable. However, this obvious relation between stability and diversity is sometimes read in reverse, and then leads to the idea that diversity promotes stability, which does not necessarily follow.

Except for the relation just noted, I see no obvious correlation between diversity and stability in the natural communities I know. Rain forests at the southern tip of South America, composed of densely packed individuals of one species of tree—a southern beech (*Nothofagus*)—with few dependent plants and animals, seem as stable as the most species-diverse tropical rain forests. And comparisons of tropical and north-temperate communities do not yield simple conclusions. The communities differ in many ways in addition to species diversity. For example, the tropical forest is not only more species-diverse but also more closely tied together—that is, cemented—by coadaptations among the species, and on the other hand it carries a heavier load of epiphytes and parasites.

The complexity and difficulty of this comparison is suggested by a (distant) analogy. Which is more stable, a pile of a few large stones, or a pile of many grains of sand? Even this simple question has no simple answer. Stability varies with different destabilizing forces: the pile of stones is more stable in wind or rain, but the pile of sand may be more stable in an earthquake. The pile of sand grains may better tolerate loss of some of its members, which may cause only limited change in the pile as a whole, while loss of one stone may bring down the whole stone pile. But, after it is dispersed, the pile of sand may be more difficult than the pile of stones to reconstitute. The pile of sand may be more effectively stabilized than the pile of stones by natural or artificial cements, which are roughly analogous to the coadaptations of organisms that hold communities together. And either the sand pile or the stone pile may be weakened by the equivalents of parasites: sand grains or stones of shapes or materials that make a pile less rather than more stable.

The stone-sand analogy *is* distant, imperfect, and oversimplified, but it at least suggests some of the difficulties in reducing the relation between species diversity and community stability to simple terms. The stability of a community may vary with different destabilizing forces, which may include changes of climate (temperature or rainfall), competition with invaders, fire, or felling by man. Some communities may tolerate loss of some of their members better than others do. But, after loss, some communities disintegrate completely while others are only modified; and some reconstitute themselves promptly while others do not. And "cement"—coadaptations—may stabilize complex communities up to a point, but may make independent existence of the species concerned impossible, and may contribute to irreversible disintegration of communities after critical losses. It is hard to see how this probable actual complexity can be adequately modeled mathematically. But it can perhaps be predicted that, in general, diverse, complexly organized communities will turn out to be relatively stable up to a critical point, but beyond this point to be less stable and more liable to irreversible disintegration than simpler communities. But this generalization, even if true, must be endlessly modified in actual cases.

Races, Species, and Higher Categories

Subspecies and Races.   As a naturalist-taxonomist I see that many species vary geographically. In some species the variation is continuous, forming clines. In other species it is more or less discontinuous; subpopulations in different places are visibly different but are (or are assumed to be) still interfertile. (For further discussion of the complexities of geographic variation, see Endler, 1977). The subpopulations are often called *subspecies,* implying that they are stages in the evolution of species, which some probably are and some not. They are usually complex within themselves, complexly interrelated, and difficult to define and distinguish precisely. But they do exist in many widely distributed species of animals and plants. They are products of the differentiation of local populations that is going on everywhere at all times, and that is one of the clearest evidences of evolution (Chapter 2).

The geographic subpopulations are often called *races.* "Race" has not only biological and evolutionary but also philosophical and ethical connotations which will be considered in Chapter 10. The point now is only that races are expected products of evolution, and that anyone who cares to look can *see* them. The Meadow Vole (which has distinguishable populations on islands off the coast of Massachusetts and New York), song sparrows, and some cychrine carabid beetles (Valentine, 1935) are a few of many visible examples.

Speciation.   *Species* may be defined, incompletely but sufficiently for present purposes, as uniquely characterized populations, or rarely single individuals, that are reproductively isolated from other populations, or nearly so.

Speciation, or the making of species by evolution, involves two processes. The first is evolutionary change in sequences of individuals or in populations. This is the essential process of organic evolution with which much of this book is concerned. The question here is: when are evolving species considered to become "new" species? And the only practical answer is: at whatever point is useful for particular purposes, which range from stratigraphic correlations to comparisons of evolutionary rates.

The second process is species-splitting, or the division of one species into two or more separate "daughter" species. The question here is: what divides populations? And the answer is different for different groups of organisms and for different modes of reproduction.

Among viruses, bacteria, blue-green algae, and asexually reproducing eucaryotes, species are difficult to define and not comparable to Mendelian species.

Among higher plants and animals that form sexually bonded Mendelian populations, the commonest or at least the most obvious way in which species split is by geographic separation of parts of populations, which permits evolution of increasingly different geographic races which eventually become species. This—*allopatric speciation*—is apparently the principal mode of species-splitting among some groups of animals, including birds and I think (from what I have seen in many parts of the world) also carabid beetles. However, most species of birds and carabids are not narrowly specialized and are rarely dependent on single species of plants or prey. Among many other animals, including specialized plant-eating and parasitic insects, and also among plants, populations may be more finely divided ecologically. Animals may also be divided by behaviors, especially mating behaviors; this is a factor in *Drosophila*, among which mating behaviors are at least partly species-specific. And populations may be divided by genetic events, including chromosome mutations and polyploidy, which may make some individuals in a population genetically incompatible with others; this probably happens more often among plants than animals. Formerly, when evolution was thought always to be a very slow process, it was difficult to see how these ecologic, behavioral, or simple genetic barriers could keep parts of populations completely separate long enough for speciation to occur, but now we see *sympatric speciation*—species-splitting without geographic separation—as supplementing allopatric speciation probably in many plants and invertebrates.

Species-splitting is too complex for further discussion here. It is well covered by other writers: Mayr's (1942) *Systematics and the Origin of Species* is a classic, and White's (1978) *Modes of Speciation* is an outstanding recent treatment.

Evolution of Higher Categories. The evolution of higher categories—genera, families, orders, phyla, and so on—is sometimes supposed to be a different process, or at least to require something more than the evolution of species. The higher categories do differ from species in that the latter are usually coherent, reproductively bonded populations with real, discoverable limits, while the higher categories, although real too in a sense (in that their members are related to each other), have no natural limits but are made wide or narrow according to the judgment of the persons who devise higher classifications and try to make them useful. (The limits and numbers of higher categories *ought* to be determined by their usefulness.) Without repeating the arguments in detail, it need be said here only that higher categories evolve from species (single species if they are clades, or more than one species if grades) by adaptation, usually species diversification, and often extinction of ancestral and intermediate forms, and that I see no need for any processes other than those involved in species evolution. Of course there is a time difference: in general, the higher the category, the longer the time required for its evolution, but this is only a general relationship (see "Generalizations about rates of evolution," Chapter 6B).

Extinction

In general, and with exceptions, extinction of species is not an isolated, deplorable event, but a natural and inevitable consequence of evolution. It compensates for the continual multiplication of species in the same way that death compensates for continual Malthusian multiplication of individuals in populations. If it were not for extinction, evolution would long ago have produced an astronomical and impossible number of species, far beyond the number the earth can hold. Darwin (*Origin*, page 109) makes this point in different words. The fossil record gives some idea of the amounts of extinction that have occurred in some groups at some times, but no more than hints at actual totals; the number of species that have become extinct is probably very many times the number that exists now.

Extinction poses problems of both fact and explanation. Massive extinctions—the disappearances of whole major groups of organisms—seem sometimes to have occurred very rapidly. Some of them may not have been quite as sudden as they seem, but they did occur, and they are difficult to explain. Attempted explanations—worldwide climatic changes, showers of cosmic rays, and episodes of continental drift—all seem more dramatic than satisfying. They leave unanswered the question, why did some groups become extinct and others not? The best guess is that mass extinctions of major groups have usually been caused by competition with new, evolving groups.

Extinction of the Dinosaurs.   The extinction of the dinosaurs at the end of the Mesozoic has intrigued both professional and amateur biologists, and various explanations have been attempted. It was not quite so sudden as popularly supposed; various groups of dinosaurs disappeared from time to time before the final episode of extinction. Of the various conceivable causes of the extinction, competition, mainly with mammals, seems the most likely. The fossil record shows that large mammals did not evolve until after the dinosaurs were gone. Competition was therefore not among the giants. But many organisms are especially vulnerable when young: plants when seedlings, young birds when they leave the nest, and so on. Young dinosaurs must have lived very different lives from the gigantic adults. They must always have had to compete with other reptiles, and diversification of small mammals at the end of the Mesozoic may have intensified competition disastrously for vulnerable dinosaur young. Even if some dinosaurs were "warm-blooded" to some extent, which is doubtful, mammals probably had evolved a more effective temperature control and therefore a competitive advantage not only in cold climates but also in many local situations elsewhere.

As for man himself—*Homo sapiens*—he is probably not threatened with extinction now and will not be until and unless a species better adapted to our exceptionally broad niche evolves on earth, from a human ancestor, or comes from another planet, which, as we learn more about the other planets, seems increasingly unlikely. The idea that insects might replace man is science fiction no less than the idea that little green men live on the other side of the moon. However, although *H. sapiens* will probably survive through the foreseeable future, it does seem likely that large populations will become extinct from time to time, most likely by the effects of atomic war, and will be replaced either by the offspring of radiation-tolerant or accidentally shielded individuals, or by populations from parts of the world where radiation was less intense.

More about extinction, and its ethical implication, will be found in Chapter 10.

Examples of Evolution at the Darwinian Level

We see adaptations—the products of Darwinian evolution by selection—everywhere around us. Striking examples that have been described recently, and that North American naturalists can see, include:

Flight of insects, birds, and man (Dalton, 1977).

Animal migration and navigation (Griffin, 1974; Johnson, 1976; Schmidt-Koenig and Keeton, 1978).

Sonar of bats and evasion of moths (Yalden and Morris, 1975; Roeder, 1967).

Coadaptations of plants and insects (Darwin, 1862; Gilbert and Raven, 1975: Patent, 1976).

Carnivorous plants (Heslop-Harrison, 1978).

Predatory mimicry of fireflies (Lloyd, 1975).

These are only a few examples, selected almost at random, of an almost infinite variety. One example that naturalists cannot see in North America is the use of electric signals by some fishes to detect obstructions and signal each other (Hopkins, 1974). It was this use of low-power electricity that pre-adapted three different groups of fishes to evolve electric organs powerful enough to stun other animals.

The following additional examples are chosen to illustrate the complexity of adaptation, the inferences that can be drawn from one case to another, and the fact that some commonplace evolutionary situations are not yet understood.

Man as a Product of Adaptation.   Naturalists should see human beings as coordinated packages of complex adaptations. The adaptive evolution of the human hand and foot, the eye, the temperature-control system, the throwing arm, the brain, and various behaviors are considered separately here (see index for pages), but these and all the other complex structural-functional systems that make up a human individual have not only evolved complexly in themselves but have also been integrated into an inconceivably complex whole which, although far from perfect (as complex adaptations must usually be) and still evolving, functions well enough in most cases. That this fact still astonishes me is, I think, not because of any inherent improbability in the complexity of the case or in the capacity of evolution to produce it, but because my brain, complex though it is, can make only oversimplified models of the still-more-complex reality.

Swifts and Swallows.   Swifts and swallows are the two principal, familiar groups of birds that make their living by feeding in flight, on flying insects, and so on, by day. They are not closely related. Technically, the swifts are not passerines but are brigaded with the hummingbirds in an order (Apodiformes) placed between (and presumably related to) the orders that include kingfishers on one hand and woodpeckers on the other, while the swallows are "oscine passerines," or songbirds. Swifts and swallows are therefore convergent groups. Both are exceptionally strong fliers and are therefore almost worldwide in distribution. Both maneuver with precision and are adept at picking insects out of the air. The swifts are the older, judging from their phylogenetic position and from the fact that they are more highly specialized both structurally and behaviorally. They spend most of their time by day and night on the wing, and cannot or do not perch as swallows do, but rest clinging to vertical surfaces. The swifts are remarkable not only in themselves but also for what can be inferred from them. Swifts are fossil in the early Tertiary; they have existed for 40 million years or so. This fact implies the *continuous* existence in the air, at least in warmer parts of the world (most swifts are tropical, and our Chimney Swift winters in South America), of an "aerial plankton" of small insects, spiders, and other small invertebrates dense enough and continuous enough to supply food for swifts for several tens of millions of years, without interruptions of more than one or a very few days. Most of the insects and spiders in the aerial plankton are dispersing as part of their normal life histories. Not only small flying insects but also small spiders enter the plankton deliberately. The (immature) spiders do it by climbing onto grass blades or other vantage points and spinning long "parachute" threads by which winds pick them up. The swifts we see, therefore, are not only remarkable examples of adaptation in themselves, but also tell us something remarkable about the adaptations of the insects and spiders they feed on. What they tell is, or ought to be, of great interest to naturalists, useful to economic entomologists and botanists working on the dispersal of insect pests and insect-carried plant diseases, and a warning to biogeographers that small organisms disperse massively through the air, so that their present distributions do not necessarily reflect old land connections or continental drift.

Tree-Trunk Feeding Winter Birds.   Bird watchers know that several of our winter birds feed entirely (brown creepers), largely (nuthatches), or partly (chickadees) on what they can find on tree trunks. These birds are somewhat differently adapted for this, in structure and searching behavior. They occupy partly different but overlapping feeding ranges (niches); their feeding ranges are probably more different than those of MacArthur's (1958) wood warblers. But, while the warblers' supply of insect food continuously renews itself during the summer, the winter tree-trunk feeders begin in the fall with a fixed amount of food and must make it last for half a year. How do they do it?

Even if only one species (population) were concerned, this question would suggest other questions rather than answers. Is the population really limited by winter food or perhaps by some other factor such as territoriality, limited nesting sites, or a density-dependent disease? (And if a disease prevents the population from becoming too dense to survive the winter on the available food, is it not beneficial, a product of coevolution?) Do the same individual birds search the same tree trunks repeatedly during the winter? I think they do. But then why do they not get most of the food the first time, and starve later? Is there a selective advantage in inefficient searching? But if so, why is it not opposed by selection in favor of individual efficiency? If involuntary restraint in use of winter food is practiced by some individuals, why do not others take advantage of it to get most of the food first?

The difficulty of this problem increases when two or more species (populations) are involved. If one exercises involuntary restraint, why does not another take advantage of it to get most of the food first? Are search patterns and the foods taken different enough to make the species effectively independent? I doubt it. The nuthatches and chickadees at least are partly opportunistic feeders, which can be seen to compete at feeding stations. In some winters northern nuthatches and chickadees make irregular migrations which may temporarily relieve local food failures. But these movements do not solve the complex evolutionary problems posed by the necessity of making winter food last while competing for it. One possibility is that the problem is insoluble; competition may end in occasional catastrophes for single species and sets of species.

These, of course, are just examples, deliberately selected because they are at hand and conspicuous, of situations that naturalists see and must try to understand.

Summary

(A) Evolution at the Darwinian level—the level that we see—has five prerequisites or components. Continuity of life, multiplication of individuals, and variation are here taken for granted or dismissed briefly. Competition and selection are defined, categorized, analyzed, and exemplified in more detail. Competition is a never-ending force propelling evolution even in stable environments. Selection is the ram of evolution, generating no force of its own but directing energy derived from replication at molecular levels. Competition and selection largely determine the force and direction of evolution, but are supplemented by genetic drift and the founder effect, and by processes at lower and higher levels.

(B) Evolution at the Darwinian level is an integrated process, partly programmed by genetic, environmental, and step-by-step requirements, but also partly open-ended. It involves "choices," selection of whole individuals in whole environments, and is too complex to be reversed precisely. Its rate varies enormously, and may be irregularly cyclical. Effective evolution apparently occurs in large, deme-divided populations, by a continuity of small steps, not by alternation of periods of equilibrium and revolution. The cost of evolution by selection—by selective elimination of great numbers of individuals—is so heavy that most populations most of the time are probably far from perfectly adapted to their environments (which are themselves evolving), but just a little better than their competitors, for the time being.

(C) Consequences of Darwinian evolution include adaptation and preadaptation, which is opportunistic rather than foresighted; specialization and focusing; adaptive radiation, parallelism, convergence, and mimicry; and diversification, including differentiation of races, species, and higher categories, all evolving by the same essential process. Species diversity and community stability are not necessarily related. Extinction is in general, and with exceptions, a natural and inevitable consequence of evolution and diversification. Examples suggest the diversity of adaptation, its complexity, and the difficulty of understanding familiar cases.

The Engine of Evolution. Evolution at the Darwinian level can be summarized in another way. What we see is the visible part of a multi-level process in which energy is transmitted from molecular levels upward through the increasingly complex sets of sets that comprise living systems. At each level sets are multiplied and diversified by processes at lower levels. And at each level differential elimination among preformed sets determines the direction in which evolutionary energy is turned. Selective forces at different levels may either supplement or oppose each other. Darwinian evolution is, therefore, comparable to an immense engine, which draws its power from molecular sources, but which moves in directions determined by a complex hierarchy of partly cooperating and partly competing, controlling and directing selective agents at different levels.

Summary of Summaries. Nevertheless, naturalists may reasonably see organic evolution simply as a process of multiplication, variation, competition, and selection, but must remember that multiplication produces hordes of excess individuals and condemns them to premature death,

that variation assures that some individuals will be genetically handicapped, that competition is ever-present, and that selection improves
populations only at the cost of elimination of enormous numbers of un-
improved individuals. Only man has (perhaps) the intelligence to escape these consequences of evolution.

READING AND REFERENCES

For naturalists, the most important book on Darwinian evolution is still the *Origin*, prefer-
ably the facsimile of the first edition, cited in Chapter 1. Other useful books are listed
with comments at the end of Chapter 1 (under *"The present"*).

SECTION A

The Continuity of Life

Farley, J. 1977. *The Spontaneous Generation Controversy from Descartes to Oparin*. Baltimore, Maryland, Johns Hopkins Univ. Press.

Variation

Wright, S. 1968-1978. *Evolution and the Genetics of Populations*. Four vols. Chicago, Illinois, Univ. Chicago Press.

Lewontin, R. C. 1974. *The Genetic Basis of Evolutionary Change*. New York, Columbia
Univ. Press.

Competition

Darlington, P. J., Jr. 1972a. Competition, Competitive Repulsion, and Coexistence. *Proc.
Nat. Acad. Sci. USA*, **69**, 3151–3155.

Wilson, E. O., and W. H. Bossert. 1971. *A Primer of Population Biology*. Stamford, Connecticut, Sinauer Associates.

De facto Coexistence of Competitors

Richards, P. W. 1969. Speciation in the Tropical Rainforest and the Concept of Niche. (In)
Lowe-McConnell (ed.), *Speciation in Tropical Environments*, New York, Academic, Pages
149–153.

The Never-Ending Role of Competition

Van Valen, L. 1973. A New Evolutionary Law. *Evolutionary Theory*, **1**, 1–30.

Selection

Darlington, P. J., Jr., 1972b. Nonmathematical Concepts of Selection, Evolutionary Energy, and Levels of Evolution. *Proc. Nat. Acad. Sci. USA*, **69**, 1239–1243.

Dobzhansky, T. 1970. *Genetics of the Evolutionary Process.* New York, Columbia Univ. Press.

Wright, S. 1969. *The Theory of Gene Frequencies.* Vol. 2 of work cited under "Variation" earlier.

Sexual Selection

See the books cited under "Bisexual reproduction" in Chapter 5.

Actual Examples of Natural Selection

Kettlewell, B. 1973. *The Evolution of Melanism.* Oxford, England, Clarendon Press.

Darlington, P. J., Jr. 1943. Carabidae of Mountains and Islands: Data on the Evolution of Isolated Faunas, and on Atrophy of Wings. *Ecological Monographs* **13**, 37–61.

SECTION B

Characteristics of the Process as a Whole

Programming of Darwinian Evolution

Darlington, P. J., Jr. 1972b. (*See* the earlier reference under "Selection.")

Irreversibility

Gould, S. J. 1970. Dollo on Dollo's Law: Irreversibility and the Status of Evolutionary Laws. *J. Hist. Biol.* **3**, 189–212.

Rates of Darwinian Evolution

Romer, A. S. 1968. *The Procession of Life.* London, Weidenfeld and Nicolson; Cleveland and New York, World Publ. Co.

Wilson, E. O. 1955. A Monographic Revision of the Ant Genus *Lasius. Bull. Museum Comparative Zool.*, **113**, 1–201. (Includes references to the Baltic amber species.)

Darlington, P. J., Jr. 1950. Paussid Beetles. *Trans. Am. Entomol. Soc.*, **76**, 47–142.

Coope, G. R. 1978. Constancy of Insect Species Versus Inconstancy of Quaternary Environments. (In) Mound, L. A. and N. Waloff (eds.), *Diversity of Insect Faunas,* Oxford, England, Blackwell Sci. Publ. for The Royal Entomol. Soc., pages 176–187.

Zimmerman, E. C. 1960. Possible Evidence of Rapid Evolution in Hawaiian Moths. *Evolution,* **14**, 137–138.

Simpson, G. G. 1944. *Tempo and Mode in Evolution.* New York, Columbia Univ. Press; facsimile reprint, 1965, New York, and London, England, Hafner.

Darlington, P. J., Jr. 1976. Rates, Patterns, and Effectiveness of Evolution in Multi-level Situations. *Proc. Nat. Acad. Sci. USA,* **73**, 1360–1364.

Modes of Evolution

Gould, S. J., and N. Eldredge. 1977. Punctuated Equilibria: The Tempo and Mode of Evolution Reconsidered. *Paleobiology,* **3**, 115–151.

MacArthur, R. H., and E. O. Wilson. 1967. *The Theory of Island Biogeography.* Princeton, New Jersey, Princeton Univ. Press.

The Cost of Selection

Haldane, J. B. S. 1957. The Cost of Natural Selection. *J. Genetics,* **55**, 511–524.

Darlington, P. J., Jr. 1977. The Cost of Evolution and the Imprecision of Adaptation. *Proc. Nat. Acad. Sci. USA,* **74**, 1647–1651.

The Imprecision of Adaptation

Van Valen, L. 1973. A New Evolutionary Law. (*See* earlier citation under "The never-ending role of competition.")

de Beer, G. 1971. *Some General Biological Principles Illustrated by the Evolution of Man,* Oxford Biology Readers, No. 1. Oxford, England, Oxford Univ. Press. (Now available from Carolina Biological Supply Co., Burlington, North Carolina 27215.)

SECTION C

Adaptation

Mimicry

Wickler, W. 1968. *Mimicry.* World Univ. Lib. New York, McGraw-Hill.

Diversification

Species Diversity and Community Stability

Goodman, D. 1975. The Theory of Diversity-stability Relationships in Ecology. *Quart. Rev. Biol.,* **50**, 237–266. (A good review of the subject, with a useful reference list; necessarily technical, but important reading for practical conservationists as well as for theoretical ecologists. Concludes "there is no simple relationship between diversity and stability in ecological systems.")

Races, Species, and Higher Categories
Subspecies and Races

Endler, J. A. 1977. *Geographic Variation, Speciation, and Clines.* Monographs in Population Biology, Vol. 10. Princeton, New Jersey, Princeton Univ. Press.

Valentine, J. M. 1935. Speciation in *Steniridia.* . . . *(See* Chapter 2, under "Snail-hunting carabid beetles.")

Speciation

Mayr, E. 1942. *Systematics and the Origin of Species.* New York, Columbia Univ. Press.

White, M. J. D. 1978. *Modes of Speciation.* San Francisco, California, Freeman. (The most important of several recent books on speciation; technical, of course. Mayr's review in *Systematic Zoology, 27,* 478–482, summarizes the contents and some conclusions.)

Extinction

Newell, N. D. 1963. Crises in the History of Life. *Scientific American,* **28,** 76–92.

Extinction of the Dinosaurs

McLoughlin, J. C. 1979. *Archosauria: a New Look at the Old Dinosaur.* New York, Viking. (Theories of dinosaur extinction are summarized on pages 101–108; see also my Chapter 8.)

Examples of Evolution at the Darwinian Level

Dalton, S. 1977. *The Miracle of Flight.* New York, McGraw-Hill.

Griffin, D. R. 1974. *Bird Migration.* New York, Dover.

Johnson, C. G. 1976. *Insect Migration.* [Oxford] Carolina Biology Readers, No. 84. Burlington, North Carolina, Carolina Biological Supply Co.

Schmidt-Koenig, K. and W. T. Keeton (eds.). 1978. *Animal Migration, Navigation, and Homing.* New York, Springer-Verlag. (Results of a symposium.)

Yalden, D. W., and P. A. Morris. 1975. *The Lives of Bats.* A Demeter Press Book. New York, Quadrangle/New York Times. (Chapter 7 is on "Seeing with Sound.")

Roeder, K. D. 1967. *Nerve Cells and Insect Behavior,* rev. ed. Cambridge, Massachusetts, Harvard Univ. Press. (A fascinating combination of technical and naturalists' writing. Chapter 5, on moths and bats, ranges from "bineural nerve responses" to "what I would do if I were a moth" hearing a bat's hunting calls.)

Darwin, C. 1862. *On the Various Contrivances by Which . . . Orchids Are Fertilized by Insects.* London, England, John Murray. Many later editions. (Of interest both in itself and as an example of Darwin's work as a serious naturalist.)

Gilbert, L. E., and P. H. Raven (eds.). 1975. *Coevolution of Animals and Plants.* Austin,

Texas, and London, England, Univ. Texas Press. (Ten contributions to a fascinating symposium.)

Patent, D. H. 1976. *Plants and Insects Together.* New York, Holiday House. (A popular but, I think, accurate small book on plant-insect relationships, many of which involve coadaptations.)

Heslop-Harrison, Y. 1978. Carnivorous Plants. *Scientific American,* **238**, 104–115.

Lloyd, J. E. 1975. Aggressive Mimicry in Photuris Fireflies: Signal Repertories by Femmes Fatales. *Science,* **187**, 452–453.

Hopkins, C. D. 1974. Electric Communication in Fish. *American Scientist,* **62**, 426–437.

Swifts and Swallows

Thomson, L. 1964. *A New Dictionary of Birds.* (*See* Chapter 2.)

Lack, D. 1956. *Swifts in a Tower.* London, England, Methuen. (A small classic.)

Tree-Trunk Feeding Winter Birds

MacArthur, R. H. 1958. Population Ecology of Some Warblers of Northeastern Coniferous Forests. *Ecology,* **39**, 599–619. (Another small classic, not directly concerned with winter birds but with resource division by competing species.)

7

ABOVE THE DARWINIAN LEVEL:
GROUP SELECTION; ALTRUISM
AND KIN SELECTION; SOCIETIES;
SETS OF SPECIES; AREA,
CLIMATE, AND EVOLUTION

The preceding chapter is concerned primarily with evolution of individuals and single populations. The present one turns to evolution of sets of individuals and sets of populations that interact in special ways. The chapter has a rough continuity: from group selection in the simplest sense; to special groups within populations (altruistic, kin, and social groups); to special sets of species (coevolving sets, faunules, faunas, and communities); to, finally, biotic or Darwin equilibriums, and the broad effects of area, climate, and numbers of species on evolution.

Group Selection

"Group selection" has been defined in several different ways and has been ignored, misunderstood, or overemphasized by different evolutionists (Darlington, 1972). I define it simply and literally as selective elimination of pre-formed groups of individual organisms (of course some groups must survive too, if evolution is to continue), or as set selection at the sets-of-individuals level. It occurs among simple groups of two or more individuals; special groups within populations; demes, which are small, local, more or less isolated subpopulations; whole species; and (in the broadest sense) groups of individuals of different species. Group selection is important in itself and doubly important because it is conspicuous in man, occurring among both geographic subpopulations and groups of individuals within populations.

Occurrence of Group Selection. A few examples of deme-group replacements have been described formally (Lewontin, 1970). Continual "overturns"—extinctions and replacements of local populations—on small islands have been recorded in detail (Simberloff and Wilson, 1970). Observant naturalists see local extinctions and replacements (as well as remarkable adaptations for survival) when ponds dry up, refill, and are repopulated by aquatic insects; they see continual small-scale extinctions and replacements of local populations of plant-eating insects in gardens; they see group selection occurring continually in man; and these are only samples of the visible evidence of group selection. Inferences suggest amounts of it. From what they see of swifts (Chapter 6C) bird watchers can infer 40 million years of insect dispersal and (by extension) of extinctions and replacements of insect populations. And from what they see of carabids, carabid watchers can infer that many species fly only to maintain themselves on unstable "ecologic checkerboards," on which local populations are continually being eliminated and replaced (Darlington, 1943, pages 43–44). The question is, not whether group selection occurs, but what its characteristics are, and what it can and cannot do.

Characteristics of Group Selection. Some evolutionists have thought of group selection as a process in which the individual is subordinated or sacrificed to the group as a whole. This idea was at least adumbrated by Darwin, Haldane, and Wright and has been emphasized and amplified by Wynne-Edwards (1962), who has argued that some group phenomena *cannot* be explained as results of individual selection, and that group selection as he defines it must be invoked as a primary creative process. Williams (1971), and others, have counterargued that Wynne-Edwards' group phenomena *can* evolve by individual selection, and that group selection is not needed to explain them. This is clearly correct. But Williams then argues that "parsimony" justifies discounting group selection, which is not correct.

Better understanding of group selection depends on understanding the relation between groups and individuals, beginning with some self-evident propositions.

Groups cannot exist without individuals. The "group as a whole" is a set of individuals, and anything that is good for the group *must* be good for one or more individuals in it. The group cannot survive unless at least one of its individuals survives and reproduces.

It is less obvious but, I think, equally true that all group selection can be converted into terms of individual selection, if concepts are precisely defined and adhered to.

A concept that requires precise definition is: fitness and selection (that is, chances of survival and reproduction) depend on the whole organism, including its whole behavior, in the whole environment, at the time fitness is measured or selection occurs. If this concept is accepted literally, as it should be in this connection, what really happens in group selection, at least in more highly organized groups, is that the fitnesses and chances of survival of some or all individuals of a group are increased in an environment that includes the other individuals. If individuals are "sacrificed," it is not to the group as a whole, but to other individuals, which then constitute the group. This way of looking at the process should remove any mystery there may be in "selection of the group as a whole."

The definition/description of group selection as selection-by-elimination of pre-formed groups of individual organisms suggests several additional self-evident propositions.

Group eliminations occur only by individual deaths (or reproductive failures), and the deaths may be spread over many generations. Among highly organized groups, deaths of a few key individuals may cause group extinctions, and in the extreme case of an idealized ant society (see under "Social groups" later in this chapter) death of a single female, the queen, can extinguish a group, but even in these cases extinctions of groups occur by deaths of individuals.

Although group extinctions occur only by individual deaths, the effect on evolution of the deaths of all individuals of a group is very different from the effect of the deaths of only some of the individuals. Group selection therefore exerts a separate and sometimes decisive evolutionary force.

Group selection has been said to be relatively slow or inefficient because group extinctions and replacements occur relatively rarely, while individual selection is a virtually continuous process. However, although selective events are relatively infrequent among groups, the effect of each event may be relatively great. For example, if a subpopulation diverges from a main population by selection of individuals during 100 generations, extinction of the subpopulation (one event in group selection) will have an effect equal to 100 generations of individual selection.

What Group Selection Can and Cannot Do. Group selection *cannot* cause evolution of group phenomena unless the latter are selectively advantageous to at least some key (reproducing) individuals in the group. However, group selection *can* do two things.

Group selection *can* oppose and counteract trends in individual selection. If, for example, individual selection during 100 generations rapidly adapts a subpopulation to a temporary local climatic cycle, at the end of the cycle group selection may remove that subpopulation, while less rapidly evolving subpopulations survive. Group selection may thus itself be a decisive homeostatic process, and may favor evolution of homeostatic mechanisms within populations; that is, populations in which genetic mechanisms favor effective evolution at not-too-rapid rates will be selected. This is an example of the principle that evolution makes situations favorable to itself.

The second thing that group selection *can* do is supplement individual selection by removing groups in which characteristics of individuals, including individual interactions, are of inferior selective value to all individuals or to key individuals. This supplementary effect of group selection is unnecessary in theory, but probably does in fact contribute significantly to the force of selection. It may be especially effective among highly organized social groups, including colonies of social insects and communities of human beings.

Altruism and Kin Selection

Biologists define altruism in several different ways. I (1978) prefer to define it broadly, as occurring in all interactions in which some individuals benefit others at cost to themselves. It may be indirect or direct, involuntary or voluntary, and one-way or reciprocal, and costs may be paid by acts or in other ways. This definition excludes concepts of volition or benevolence, and includes much that is not altruism in the everyday meaning of the word; the narrower, everyday concept of altruism will be related to it.

Altruism is a group phenomenon. It requires at least a group of two, an altruist that pays a cost and a recipient that receives a benefit. Such groups do form themselves, vary, sometimes compete, and evolve by selection. This is a special case of evolution by group selection, and it has the same characteristics as other cases: groups must be formed, usually by action of their members, before selection can occur; selection is not by positive choice, but always by negative elimination; and selection among altruistic groups, as among other groups, can always be converted into terms of selection of individuals in environments that include other individuals.

Genes and Individuals, Selfish and Unselfish. Genes and the individuals carrying them must be selfish. Any kind of gene that consistently sacrificed itself to the benefit of other kinds, with no chance of any return, would eliminate itself from its gene pool. Unselfish, one-way "genes for altruism" cannot exist, although they have been invoked many times by many biologists. Dawkins (1976) makes this point at book-length in *The Selfish Gene,* and other persons have said the same thing in fewer words. It follows that altruistic individuals that do not receive or "expect" any kind of return, cannot be genetically differentiated within populations of selfish non-altruists. Exceptions to this generalization are so trivial that they do not seriously impair it.

The preceding generalization applies to cases in which altruistic relationships are initiated by altruists, but in other cases comparable relationships are initiated by recipients; some genes or individuals take benefits from others by actions equivalent to predation or parasitism. This is not altruism as usually defined, but it is difficult to distinguish from altruism if concepts of volition or benevolence are excluded. It can be thought of as *forced altruism,* and as such it is important for two reasons. It is a very common one-way relationship in the real world, and it may often be the first stage in evolution of reciprocal altruism, as suggested under "Evolution of altruism," which follows later in this chapter. However, "forced altruists" are not unselfish; they pay the costs imposed by predation or parasitism because they must, and if they can.

Although, with exceptions that do not seriously impair the generalization, genes and the individuals carrying them must be presumed to be selfish, they often behave in ways that *as separate events* are unselfishly altruistic; they "benefit others at cost to themselves." But there are ways in which the presumption of selfishness and the fact of altruism can be reconciled.

Kin Selection.   Kin selection reconciles the presumption of selfishness with the fact of altruism. In it, individuals sacrifice themselves to their kin altruistically, but "expect" selfish genetic profits. The theory is important because it is often applied to man.

The theory is based on simple arithmetic. In ordinary, sexually reproducing populations, two nonidentical sibs (brothers and sisters) are said to inherit one-half of the same genes; an uncle or aunt and a nephew or niece, one-fourth; two first cousins, one-eighth; and so on. These genes are identical, derived by replication of the genes of most-recent common ancestors. From these fractions it is commonly calculated that an altruist will profit genetically—will "increase the representation of its genes in the next generation"—if it sacrifices itself for more than two sibs, more than four nephews or nieces, or more than eight first cousins; or if it makes a partial sacrifice of less than one-half of its "personal fitness" for a sib, less than one-fourth for a nephew or niece, and so on.

An analogy clarifies this calculation. When a sum of money is given by one group of persons to another, it makes no difference to the recipients whether the sum is made up by direct gifts from all givers, or whether some individuals give indirectly, through others; and if indirection increases the sum given, it may be preferred. Similarly, if a set of genes is given by one generation to the next, it makes no difference to the next whether the set is made up by reproduction of all givers, or whether some individuals give indirectly, by altruistically increasing the reproduction of others; and if indirection increases the sum of genes given, it may be "preferred" by natural selection.

However, the following factors complicate the arithmetic and modify or reverse the conclusions usually drawn from it:

1. In ordinary, sexually reproducing populations all individuals share many genes. Kin therefore share many more genes than the fractions show. But non-kin share many genes too, and this greatly reduces the genetic "profits" of altruism to kin, as compared with altruism to non-kin. If all individuals in a population share all their genes, as in some asexually reproducing clones, altruism to kin is no more profitable than to non-kin; an individual that sacrifices itself to any other in the population hands on a set of genes equivalent to its own. And if, as geneticists think (for example, C. D. Darlington, 1978, page 71), most human individuals share most of their genes, with only a small fraction of the genes being significantly different, altruism to kin is only a small fraction more profitable than to non-kin. Some genes shared by non-kin may vary molecularly in ways that are selectively neutral (see Chapter 5, "The Selectionist-neutralist controversy), but this should not affect the calculation. In the preceding analogy, it does not matter to recipients whether they receive dollars or other currency, if the purchasing power is the same, and it should not matter to the recipients of genes whether the latter are identical or molecular variants, if the selective effects are the same.

2. Kin altruism yields the calculated genetic profits only if it makes the whole difference between recipients producing no offspring and producing full quotas. It is proportionately less profitable if recipients have already produced fractions of their quotas, or if they have only fractional chances of reproducing. For example, altruism to young kin that have only 1 chance in 20 of surviving later risks and reproducing will yield only $\frac{1}{20}$ of the calculated genetic profits. It follows that kin altruism should be most profita-

ble when recipients are young adults just ready to reproduce. But this is just when most animals are most strongly attracted to *non-kin*.

The two preceding paragraphs suggest that the arithmetic of kin selection is too simple. And the simple model may be further complicated, and predictions made from it greatly changed, by additional factors, including the following:

3. Difficulty of kin recognition at the most profitable time, among young adults. (See the preceding discussion.)

4. The cost of evolution by selection (selective eliminations) of complex kin recognition and preference behaviors, which must compete with a great number of other selective processes for payment of costs. Costs may limit or inhibit the evolution even of selectively advantageous behaviors. (See Chapter 6B, "The cost of selection.")

5. Competition also for space on the brain, which is crowded with other behavior-determining mechanisms. This too may limit or inhibit evolution of selectively advantageous behaviors. (Chapter 6C, "Specialization and focusing.")

6. The cost of suppressing genetically determined behaviors, maintained by selection since the beginning of life, that favor an individual's own offspring.

7. Kin competition.

8. The possibly greater profits of non-kin altruism.

The last two items require explanation. An individual cannot at once both altruistically increase and competitively decrease another's chances of surviving and reproducing, and when competition is more profitable, it should inhibit kin altruism. For example, if one brother eliminates another and reproduces in his place, the one hands on twice as many of his genes as he would by sacrificing himself to his brother. Brothers should therefore be under intense compulsion to replace each other. But they are not. Failure of this simple prediction should warn us to look critically at the more complex and less profitable arithmetic of kin selection. Kin selectionists would probably say that the advantage of inhibiting a brother's reproduction and maximizing one's own is overridden by other factors in complex situations. This is surely true. But the most profitable alternative to brothers' competition may usually be, not kin-oriented altruism, but non-kin reciprocal altruism.

Kin-oriented altruism requires complex behaviors of kin recognition and kin preference; an individual helps and can expect help from only a few other individuals and must, in effect, discriminate against all the rest; and kin are likely to be scattered out of reach of each other after dispersal. Non-kin reciprocal altruism is simpler; an individual can expect to be helped by many others and discriminated against by none, and is more likely to find altruists within reach after dispersal. Non-kin reciprocal altruism should therefore be much more profitable than kin-oriented altruism, especially in populations in which all individuals share many genes and in which dispersal occurs. Even in these cases, close kin will probably help each other, but they will do so, not primarily because they are kin, but because they occur together and are often (but not always!) compatible.

The preceding discussion is concerned with the *theory* of kin selection. To test it, naturalists should *look* at real cases, and especially at real people. Kin selection theory predicts that brothers will be altruistic to each other for their genes' sake, whether they like each other or not. But look at Cain and Abel. Cain was not a real person, but he was a paradigm of what people saw as reality. The histories of Roman emperors and English kings offer real examples of brothers being less than altruistic to each other or to their kin, and less exalted brothers can behave in similar ways. Brothers are *not* always altruistic to each other for their genes' sake. Anthropologists (including Sahlins, 1976) see that altruism in primitive societies is *not* precisely determined by kinship. And non-kin friends can be as mutually altruistic as brothers. I think that naturalists who look objectively at real people will probably decide that what they see does *not* fit the predictions of kin-selection theory, but does fit a theory of reciprocal altruism in which kin are included but not uniquely favored.

The conclusion is that in most cases competition does inhibit altruism; that when it does not, reciprocal altruism among all responsive individuals in populations is simpler and more profitable than kin-oriented altruism; and that kin behave altruistically to each other, not because they are kin, but because they occur together in populations of reciprocally altruistic individuals.

Kin selection does occur in two special cases, in which it is so conspicuous that it emphasizes, by contrast, how little evidence of it there is in other cases. One is parent-offspring groups, in which female and sometimes male parents behave altruistically to their offspring, but benefit their own genes. The other is colonies of ants and other social Hymenoptera, in which female workers are half-identical twins, which greatly increases the genetic profits of their kin altruism. (See the discussion of the idealized ant colony later in this chapter under "Examples of altruism" and "Social groups.")

Kin selection is treated in detail here because it has been overemphasized recently. Most evolutionists, even some who are not otherwise determinists (Chapter 10), do accept it uncritically. And there is nothing wrong with its simple arithmetic, as a preliminary model. The trouble is that in real situations the simple model is modified by many factors that change or reverse predictions made from it, and that when we look for it in the real world, we do not see it, except in very special cases.

The literature on this subject includes a long-overlooked paper by Snell (1932) on sex determination and kin altruism in Hymenoptera, a recent "critique" by Grant (1978), and a recent "commentary" by Schulman (1978). The last two papers list other recent works.

Reciprocal altruism. The presumption of selfishness of genes and individuals and the fact of altruism are reconciled also in reciprocal altruism (Trivers, 1971), in which two or more individuals (or other entities) benefit each other regardless of kinship, each profiting more than its altruism costs. In contrast to kin altruism, which is clearly visible only in exceptional cases, reciprocal altruism is ubiquitous in the real world. It occurs or may occur between "copies" of the same gene, between different genes in a gene pool, between genetically similar and between genetically different individuals in a population, and between individuals of different species.

Four Propositions about Altruism. Four propositions about altruism follow, directly or indirectly, from the preceding discussion. They cannot be proved, but can be tested, are consistent with each other, and can be diversely exemplified. They are apparently true of altruism in kin groups as well as in other situations, but not of "forced altruism."

1. Altruism is always actually or potentially *reciprocal;* altruists always receive or "expect" (are statistically likely to receive) direct or indirect returns.

2. Altruism is always actually or potentially *profitable* to all individuals concerned; all individuals at the beginning of their existence (when the eggs are fertilized or even before) "expect" to gain more than they lose, although the expected profits of different individuals need not be equal.

3. Altruists and recipients are always *environmentally,* not genetically, differentiated, usually by *position* in kin groups or environmental situations.

4. In summary, altruism is a *net-gain lottery* in which all individuals pay or risk paying costs, but receive or "expect" profits that exceed the costs.

These propositions are apparently true of all cases in which altruism is initiated by the altruists and is evolving or is stabilized by selection. Theoretical exceptions do not seriously impair the propositions. "Forced altruism," which is not initiated by the altruists, does not conform to the propositions but probably tends to evolve in directions suggested by them.

Examples of Altruism. The following somewhat arbitrary, oversimplified examples illustrate the preceding propositions.

Mendelian populations produce surplus individuals, which are essential to allow normal losses and gene substitutions, and which are altruists by definition; by dying or failing to reproduce they pay costs that may be thought of as benefiting the populations as wholes but actually benefit surviving individuals. The altruism is potentially reciprocal; any individual may be either an altruist or a recipient. It is potentially profitable to all individuals; it assures survival of some individuals, and all individuals have a chance to be the survivors. And altruists and recipients are environmentally determined, by position in group or environmental situations—for example by which individuals happen to meet predators. This is a net-gain lottery in relatively simple form.

In kin groups too altruism is potentially reciprocal; any individual may be a sib, or an uncle (or aunt) or a nephew (or niece), or a cousin, and may pay costs or receive benefits, or both, accordingly. It is potentially profitable, directly or indirectly to all individuals concerned. And altruists and recipients are determined by position in kin groups and in environmental situations, according to which individuals happen to be in position to give most or profit most, or to need help or give it. This is a somewhat more complex net-gain lottery.

In an idealized ant colony altruism is more complexly reciprocal; female parents (queens) produce eggs and rear the first offspring altruistically, and some female offspring (workers) then reciprocate the altruism, caring for their parent at the cost of their own sterility, and workers behave altruistically to each other. The altruism is potentially profitable, directly or indirectly, to all females; all, at the beginning of their lives, have a chance of becoming queens and benefiting accordingly, and even the workers "increase the representation of their genes in the next generation" by their altruism. The "expected" profits of a female are then its chance of becoming a queen plus the indirect genetic profits of its altruism to close kin; the sum would not be easy to calculate! And altruists and recipients—worker castes and queens—are environmentally determined. This is a still more complex net-gain lottery. (For further analysis of this example, see "Social groups," discussed later.)

The ontogeny of a multicellular organism exemplifies the propositions at another level. The somatic cells are altruists, the reproductive cells recipients; and the somatic cells interact altruistically among themselves. The altruism is potentially reciprocal; at least early in ontogeny any cell may become either a reproductive or a somatic cell or may change its roll among somatic cells. All the cells presumably profit from their altruism; they all carry the same genes, and their altruism increases the chances that their genes will be represented in the next generation. And altruists and recipients—the somatic and reproductive cells—are differentiated primarily by position in the early stages of ontogeny. This too is a net-gain lottery, simple in theory but very complex in fact.

Real cases are more complex and difficult than these oversimplified theoretical examples. This fact is illustrated by a case previously considered by several writers including Trivers (1971, page 43; he gives additional references). If one bird in a flock calls an alarm when it sees a hawk, risking itself but warning the others, the act, seen as a single event, looks like one-way altruism. However, if, because of the alarm, the hawk catches no bird and goes away to hunt easier prey, the alarm-caller may profit from its own act; the calling then looks less altruistic. Or if later events show that any bird in the flock may either call an alarm or receive a warning, the altruism is seen to be not one-way but reciprocal. A careful observer might see all this. However, if the flock includes close kin, kin-group effects that appear only in the next generation may modify the immediately visible effects, and would be very difficult to see.

Evolution of Altruism. Altruism is a group phenomenon and evolves by group selection, which, as noted earlier, can always be converted into terms of individual selection. If it is asked how groups that consist of two or more individuals playing different roles can evolve by individual selection, the answer is by coevolution or its equivalent.

One gene or one kind of individual may (in effect) coevolve with itself if "copies" of it play different roles in different positions in kin groups or environmental situations; it is the roles that (in effect) coevolve. A gene that induces altruism among close kin might do this, and coevolution without genetic differentiation of altruists and recipients may occur also in responsive altruism. This might begin with appearance of a mutant gene which causes the individual carrying it to offer responsive altruism to other individuals. If the gene does not appear in other individuals, the offers will be refused and the gene will probably be eliminated. However, if, by the chances of Mendelian heredity, the F_1 or F_2 includes more than one individual carrying the gene, offers of altruism may elicit altruistic acts in response; responsive altruism may become selectively advantageous; and a behavioral group may be formed and may evolve. This does not require spatial isolation, and, although it is most likely to begin among close kin, it does not depend on kin selection. Groups like this are conspicuous in man.

Coevolution of genetically differentiated individuals within a population is exemplified by evolution of reciprocally altruistic males and females in Mendelian populations.

And coevolutions of reciprocally altruistic species range from sets of two, as the fungus and alga in a lichen, to sets of many species, as in complexly organized communities.

Coevolution of reciprocal altruists may be inhibited by "cheaters," which accept benefits from altruists but make no return (Trivers, page 44), but reciprocal altruism is in fact so common that cheating must often be avoided, countered, tolerated, or used. In kin groups, effective cheating might require mimicking kinship, which might be difficult. In responsive altruism, the required response might be an actual altruistic act, which could not be mimicked. Or in an intelligent animal, perhaps only in man, cheating may be recognized and countered. However, these are special cases. A more general case may begin with a move, not by an altruist, but by a recipient, which may initiate a simple host-parasite or prey-predator relationship, the host being a "forced altruist." Host and parasite may then coevolve in ways advantageous to themselves, the host toward tolerance for the parasite, the parasite toward moderation in its demand on the host because outright elimination of the host would eliminate the parasite too. Finally, the host may "use" the parasite, perhaps as a source of vitamins, or as a means of eliminating weak or overage individuals, or directly in defense against other parasites, or indirectly as a "weapon of competition." The relationship will then have evolved into reciprocal altruism. This kind of coevolution, endlessly diverse in detail, may be a multi-level process, occurring among genes and individuals in single populations as well as among different species.

Three general comments can be made about evolution of altruism as follows:

1. Evolution of altruism is always opposed by competition; an individual cannot at the same time both altruistically increase and competitively decrease another's chances of surviving and reproducing. Altruism will not evolve unless it is more profitable than competition.

2. The kinds and amounts of altruism that are selectively advantageous depend on complex interactions in groups that usually include both kin and non-kin individuals, and the proportions of kin and non-kin, and opportunities for interactions, change in dispersal rates and population densities, which themselves change with amounts and distributions of resources. Change in one resource may greatly change the kind and amount of altruism that is profitable in a particular case and may also change the kind and amount of opposing competition. This is only one of many possible examples of how environmental change may change patterns of altruism. The patterns are presumably unstable; and costly, slow natural selection should not be expected to produce complexly precise behaviors to fit them. A generalized, flexible altruism may be both simpler in its evolution and more advantageous.

3. Altruistic-group selection is presumably supplemented by group selection at a higher level. Groups in which altruism is relatively ineffective are presumably eliminated, not by a mysterious groups-as-wholes effect, but by differential elimination of individuals that do not receive the reciprocal advantages of altruism. An example is suggested under "Altruism in man," which follows next.

Altruism in Man.   Analysis of altruism in man can raise unscientific emotions even among biologists. I shall try to ask questions and offer tentative answers rather than to pontificate about it.

First, does analysis in terms of genes and selection and costs and profits "take the altruism out of (human) altruism?" I think not. Humans do not calculate the profits of altruism, but pay the costs or run the risks of it because they have human altruistic emotions.

Does altruism in man conform to the four propositions offered earlier in this chapter? I think it does; at least this is a good working hypothesis. Human altruism does seem to be potentially reciprocal and profitable; some individuals do happen to lose by it, but all "expect" (are statistically likely to receive) profits that exceed their costs, although the expectation is usually not conscious, and although the profits are (and are expected to be) unequal. And altruists and recipients are environmentally determined, by position in groups or environmental situations;

no persons are condemned by their genes to be altruists with no chance of being recipients. Altruism is therefore a net-gain lottery for humans as for other organisms.

Is altruism in man genetically determined? I think it is and is not. A man is a man and behaves as one because of his genes. However, no specific "genes for altruism" have been found in man: human altruism seems more general and flexible than rigidly predetermined.

What part do kin altruism and kin-group selection play in man? Parent-offspring altruism is conspicuous and its importance unquestioned. Altruism is conspicuous also among human sibs and less-close kin, but it is not precisely proportional to kinship; it is greatly modified, either reinforced or cancelled, by individual likes and dislikes. And among young adults, among whom kin altruism should be most profitable genetically, it is overridden by attractions and reciprocal altruism between *non-kin* males and females. In general, altruism in man seems correlated with congruity and compatibility more than with kinship, and this suggests that human altruism is responsive more than kin-related. Sahlins (1976) makes the same point—that human altruism is not closely correlated with genetic kinship—in greater detail; readers of his book should read also Simpson's review of it, cited in my reference list.

How has altruism evolved in man? Altruistic behaviors do not fossilize, and evidence derived from observation of surviving hunter-gatherers and nonhuman primates is indirect and incomplete, so the answer must be largely inference and outright guessing. We can guess that among our remote prehuman ancestors altruistic behaviors may have been determined in some detail by specific genes and may have evolved by genetic variation and selection, but now, although our altruism still has a broad genetic base, the details of it and its evolution seem to be determined socially more than genetically. This has involved a shift to evolution at a new, social level. Variation and selection occur at the new level, but with new characteristics: the variation is amplified by something like inheritance of acquired characters, which greatly increases the rates and amounts of diversification that occur before selection; and, although selection is still primarily by elimination, it can be supplemented by intelligent choice. Diversification and evolution at the new level are increasingly rapid and complex, but still, I think result in coevolutions of reciprocal altruists.

Does altruism in man conform to the general comments just made? I think so. Human altruism is opposed and sometimes inhibited by competition. It is imprecise in detail; some biologists seem to suggest that evolution by selection has produced altruistic behaviors in humans precisely adapted to degrees of kinship, the costs and profits of kin and non-kin altruism, and the profits of selfish cheating and the counter-profits of detecting and preventing it. But we have only to look around us to see that human altruism is not so precisely detailed. It is also not precisely limited; our altruistic emotions often extend beyond the limits of man to cats and whales and even plants. (However, reason may justify man's altruism to these and other organisms.) And group selection does supplement individual selection in evolution of altruism in humans; this is illustrated by control of "cheating."

"Cheating," or acceptance of altruism without reciprocation, does occur in human populations. Milder forms of it, including minor failures of responses or cooperation, are recognized and countered or tolerated by individuals. More serious forms are controlled by laws or other group actions. However, control is not perfect. Human populations usually carry loads of non-altruists, and if the loads become too great, whole populations may be eliminated; this results in a kind of group selection that has presumably continually supplemented individual selection in evolution of altruism in humans, and may still operate. "Forced altruism," including outright slavery, also occurs in man, and does sometimes evolve into reciprocal altruism.

My tentative and I hope not dogmatic conclusion is that altruism as we see it in ourselves—the narrower everyday concept of it—fits remarkably well into the broader biological concept.

Social Groups

E. O. Wilson, in *The Insect Societies* (1971) and *Sociobiology* (1975), has brought together an enormous amount of information about the organization, behavior, and evolution of animal societies. These books are fascinating for naturalists, who should go to them for details. But note that the last chapter of *Sociobiology*, which concerns man, and which has been unmercifully damned by some reviewers (see my Chapter 10), is superseded by Wilson's more recent *On Human Nature* (1978), which has received a Pulitzer prize.

Wilson defines a society as "a group of individuals belonging to the same species and organized in a cooperative manner . . . transcending mere sexual activity." By this definition societies are groups of reciprocally altruistic individuals, and they presumably evolve by group selection, modified by altruistic behaviors, and by kin effects when the groups are composed of close kin.

An oversimplified model of an ant colony—a simple society—is worth analysis for its own sake and for what can be inferred from it. (The analysis is not Wilson's but from a 1972 paper of mine cited earlier in this chapter.) The colony consists of a female parent, the queen, and her female offspring, the workers. The latter are half-identical twins, which have all received the same genes from their haploid father (see again Snell, 1932). The ontogeny of the colony (as well as its phylogeny) requires three genes or sets of genes, all of which are present in all females: one keeps the female parent and her female offspring together and induces reciprocal altruism among them; a second makes all females susceptible to becoming either queens or workers, depending on the food they receive; and a third induces special behaviors in special situations.

This is a tri-determinant genetic system, like the system that determines the ontogeny of a multicellular individual organism (see under "Examples of altruism" earlier). An ant colony is similar to a multicellular organism in other ways; both are organizations of genetically identical or half-identical, reciprocally altruistic units. The society is therefore analogous to a super (supra) organism. This close analogy (see Chapter 3) was fashionable once, was then disparaged as "not heuristic," and is now becoming acceptable again. It should be an effective explanatory and teaching analogy, which should help naturalists understand what they see in an ant colony, and which allows important predictions and inferences. It would have allowed the prediction that castes of social Hymenoptera—queens and workers, and different kinds of workers—are differentiated by environmental factors rather than genetically; that this is indeed the rule is only now being accepted by entomologists, after a century of argument (Oster and Wilson, 1978).

The diverse behaviors of ants in a colony include the queen's marriage flight and mating, nest founding, laying of unfertilized and fertilized eggs, and initial care of eggs and larvae; and the workers' care of the queen and brood, nest building and nest cleaning, defense, foraging including finding food and leading other workers to it, and the reciprocal feeding and care of each other. These activities are partly shared and partly divided. The queen and the workers play largely different roles, but queens behave briefly as workers, and workers sometimes function as supplementary queens. And individual workers share some activities and divide others. All this behavior is genetically determined; we know this because different species of ants inherit different "behavioral repertories." But what individual ants do within their species' repertory is determined nongenetically, by the kind or amount of food they receive as larvae (which differentiates queens and workers, and major and minor workers) and by their positions in social interactions and group activities. So, the (involuntary) decision as to which of a variety of behaviors an ant will perform is not genetic, but after the decision is made the ant's behavior is precisely gene-determined. This case will be referred to again in discussion of determinism in Chapter 10.

Sets of Species

This heading covers all sets of two or more interacting species: sets of simple resource competitors; host-parasite sets, including prey and predators, and plants and the animals that eat them; coaltruistic or symbiotic sets as various as fungi and algae in lichens (Smith, 1973), flowering plants and their insect pollinators (Richards, 1978, with references to earlier work including Darwin's), the members of Müllerian mimetic associations (Chapter 6C), ants and their aphid "cows," and men and dogs; and the more-complex sets of sets of species that make up communities and biotas. All these sets evolve by coevolutions (coadaptations and counteradaptations) of the different species involved, supplemented by group selection—that is, by the differential elimination (and survival) of groups as wholes, which can always be converted to individual selections. The coevolutions of all these interacting groups form an inextricably, inconceivably complex network over the earth. Naturalists can see only fragments of it clearly, but may see, less clearly, some characteristics of the whole (as Darwin did), and may contribute to the understanding of it (as, again, Darwin did).

Formation and Evolution of a Faunule. The process of formation and evolution of small sets of species in small, isolated areas has been observed on small islands (MacArthur and Wilson, 1967) and can be seen on the arctic-alpine summits of the Presidential Range (Mt. Washington) in New Hampshire (Darlington, 1943). Here, a small "faunule" of pre-existing (*not* locally evolved) species of Carabidae, chiefly but not exclusively flightless, ground-living species pre-adapted to the arctic-alpine climate, has accumulated, while other, flying species that continually reach the mountain tops are continually eliminated, as described in Chapter 6A under "Actual examples of selection in nature." This faunule has been formed by selection among species, and it will evolve by further species selection, notably by selective elimination of species associated with post-Pleistocene ponds and bogs that are gradually being drained by erosion. It is easily accessible, and naturalists who know a little about Carabidae can see it for themselves and perhaps contribute to knowledge of it by recording additions and losses of species from time to time.

Darwin Equilibriums.   Darwin (*Origin,* page 109) said:

> . . . as from the high geometrical powers of increase of all organic beings,
> each area is already fully stocked with inhabitants, it follows that as each
> selected and favored form increases in number [of individuals], so will the
> less favored forms decrease . . . but we may go further than this; for as
> new forms are continually and slowly being produced, unless we believe
> that the number [of species] goes on perpetually and almost indefinitely
> increasing, numbers must inevitably become extinct.

This is the first statement of the relation between additions and sub-
tractions of species that produces biotic equilibriums. Because Darwin
did state this relation, even though imprecisely, and in order to emphasize the continuity of this and many other concepts from the
premathematical era of biology to the present, I (1972, page 3152) have
proposed that the equilibriums that determine the numbers of individuals and of species in "each area" be called "Darwin equilibriums."

MacArthur and Wilson (cited earlier), and others, have now begun to
develop mathematical models for analyzing and predicting these equilibriums. However, although the theory can be treated with mathematical precision, the different interactions that determine equilibriums in
actual cases cannot yet be sorted out, except in artificially simple situations. We can count the numbers of individuals and of species in biotas,
and relate the numbers to area, distance, and other factors, but we cannot yet separate the parts played by competition in the narrow sense,
predation, and disease, and by cooperation and coevolution in determining the numbers. The concepts of extended competition and competitive repulsion resulting from a totality of interactions cover cases
like these.

Area, Climate, and Evolution

The inconceivably complex network of interacting, evolving sets of species, and the competitive repulsions and Darwin equilibriums they generate form an apparent worldwide pattern of evolution of dominant
plants and animals in the largest and most favorable areas and dispersal
into smaller and less favorable ones. Darwin saw parts of this pattern
and inferred the whole. He said (*Origin,* pages 337–338):

> From the extraordinary manner in which European productions have re-
> cently spread over New Zealand, and have seized on places which must
> have been previously occupied, we may believe, if all the animals and
> plants of Great Britain were set free in New Zealand, that in the course of

time a multitude of British forms would become thoroughly naturalized there and would exterminate many of the natives. On the other hand, from what we see now occurring in New Zealand, and from hardly a single inhabitant of the southern hemisphere having become wild in any part of Europe, we may doubt, if all the productions of New Zealand were set free in Great Britain, whether any considerable number would be enabled to seize on places now occupied by our native plants and animals. Under this point of view, the productions of Great Britain [which are those of Eurasia, of which Great Britain is a recent continental island] may be said to be higher [more dominant] than those of New Zealand. Yet the most skillful naturalist from an examination of the species of the two countries could not have foreseen this result.

Much more of this kind of evidence can be seen now. Man, intentionally or accidentally, has carried many plants and animals, especially weeds and insects but also some vertebrates including the House Mouse, Black and Norway Rats, and the English Sparrow, back and forth in every direction. But man-carried species have taken hold more in some directions than in others. Species from Eurasia have probably established themselves in North America more than the reverse; from Eurasia and North America, in Australia and probably also in southern South America more than the reverse; and from all the continents, on islands more than the reverse.

Dispersals of animals (and plants?) not carried by man apparently form a comparable pattern. The most dominant (successful) groups of vertebrates seem usually to have evolved in the huge area and favorable climate of the Old-World tropics and spread from there, each new group tending to replace older ones as it spread. Within this main pattern, mammals seem continually to have moved (spread) from Eurasia to North America across successive Bering land bridges more than the reverse, and birds too seem to have moved mainly in this direction. Many North American mammals moved (spread) into South America over the Late Pliocene land connection, with countermovements of South American mammals fewer and mostly temporary. Many recent Asiatic vertebrates have moved toward and to Australia, while older Australian groups have made only a few, small countermovements toward Asia. And vertebrates everywhere have moved much more from continents to islands than the reverse. All this is described in more detail in my *Zoogeography* (1957, especially pages 545–547). Invertebrates are less well known, but the direction of movement of Carabidae, at least, seems to have been from Asia toward Australia more than the reverse (Darlington, 1971) and from continents to islands.

The broad pattern that Darwin partly saw and partly inferred he sum-

marized by an effective analogy (*Origin*, page 382). He compared evolving and dispersing plants and animals to "... living waters ... that have
flowed with greater force from the north so as to have freely inundated
the south," and I (1959, pages 314–315) have rephrased and extended
his analogy, saying:

> We know ... that Darwin's "living waters," the invisible rivers that living
> things make as they evolve and flow over the earth, are not random but
> have a central pattern ... This river analogy is too simple. The invisible riv-
> ers that evolving, moving animals make do not flow directly but are like
> tidal rivers that flow back and forth (but still have a net direction) and are
> so full of complex cross currents and whirlpools that the main flow is hard
> to see. But underneath the details there is the main pattern that Darwin
> saw and that we see more clearly now: evolution of successive dominant
> groups in the largest favorable areas and movement (spreading) to smaller
> areas, from the great continents to smaller continents, and from all the
> continents to islands [and from favorable to unfavorable climates].

Darwin's explanation (*Origin,* page 379) was:

> I suspect that this preponderant migration from north to south is due to
> the greater extent of land in the north, and to the northern forms having
> existed in their own homes in greater numbers, and having consequently
> been advanced through natural selection and competition to a higher
> stage of perfection or dominating power than the southern forms ...

And he used another effective analogy (page 380): "... the more dominant forms, generated in the larger areas and more efficient workshops
of the north." In other words (my words) the worldwide pattern is generated by evolution of dominant groups in large favorable areas and
their spread from there; and they evolve there because large numbers of
species and intense competition induce effective evolution by selection,
selection occurring (I think) among individuals, sets of individuals,
whole species, and sets of species. This pattern conforms to the geographic-climatic competition gradient summarized in Chapter 6A (under
"The distribution of competition in space and time"). Dobzhansky
(1950) suggests (in effect) that selection in unfavorable climates makes
plants and animals climate-tolerant, while selection in more favorable
places, where species are more numerous and competition more intense, makes them better competitors. And I think that the cost (Chapter 6B) may forbid effective selection in both directions at once.

All this is hard to see and partly speculative. Many evolutionists do
not see it and are not interested in it; their minds are focused on molecules, not worldwide patterns formed by living things. The molecules

are important, but the wider patterns are important too. They are products of evolution, and can tell us things about evolution in the real world that molecules cannot.

Summary

Group selection, defined as selective elimination of pre-formed groups of individuals, is not a process in which individuals are sacrificed to the group as a whole, but one in which some individuals are sacrificed to other individuals, which then constitute the group. In such groups, altruism is a net-gain lottery, and kin selection is less profitable than reciprocal altruism except in special cases; reciprocal altruism may evolve from host-parasite or prey-preditor relationships. Social groups are tri-determinant organizations of reciprocal altruists. Groups or sets of species evolve by selection as faunules and faunas which form Darwin equilibriums; over the world as a whole they form an inconceivably complex system in which dominant groups tend to evolve in the largest, most favorable areas and disperse to smaller, less favorable ones.

READING AND REFERENCES

Group Selection

Darlington, P. J., Jr. 1972; 1975; 1978. Papers in *Proc. Nat. Acad. Sci. USA,* **69**, 293–297; **72**, 3748–3752; **75**, 385–389.

Lewontin, R. C. 1970. The Units of Selection. *Ann. Rev. Ecol. and Systematics,* **1**, 1–18.

Simberloff, D. S., and E. O. Wilson. 1970. Experimental Zoogeography of Islands. A Two-Year Record of Colonization. *Ecology,* **51**, 934–937.

Darlington, P. J., Jr. 1943. Carabidae of Mountains and Islands: Data on the Evolution of Isolated Faunas, and on Atrophy of Wings, *Ecological Monographs,* **13**, 36–61.

Wynne-Edwards, V. C. 1962. *Animal Dispersion in Relation to Social Behavior.* Edinburgh, Scotland and London, England, Oliver and Boyd.

Williams, G. C. (ed.). 1971. *Group Selection.* Chicago, Illinois, Aldine-Atherton.

Wade, M. J. 1978. A Critical Review of the Models of Group Selection. *Quart. Rev. Biol.,* **53**, 101–114.

Altruism and Kin Selection

Darlington, P. J., Jr. 1978. Altruism: Its Characteristics and Evolution. *Proc. Nat. Acad. Sci. USA,* **75**, 385–389.

Genes and Individuals, Selfish and Unselfish

Dawkins, R. 1976. *The Selfish Gene.* Oxford, England, Oxford Univ. Press.

Kin Selection

Darlington, C. D. 1978. *The Little Universe of Man.* London, England, George Allen & Unwin.

Sahlins, M. 1976. *The Use and Abuse of Biology. An Anthropological Critique of Sociobiology.* Ann Arbor, Michigan, Univ. Michigan Press. (Readers should *see also* Simpson's review, *Science,* **195**, 773–774.)

Snell, G. D. 1932. The Role of Male Parthenogenesis in the Evolution of the Social Hymenoptera. *American Naturalist,* **66**, 381–384.

Grant, V. 1978. Kin Selection: A Critique. *Biologisches Zentralblat,* **97**, 385–392. (Not widely available, but reprints can be requested from the author, Department of Botany, University of Texas, Austin, Texas 78712.)

Schulman, S. R. 1978. Kin Selection, Reciprocal Altruism, and the Principle of Maximization: A Reply to Sahlins. *Quart. Rev. Biol.,* **53**, 283–286.

Reciprocal Altruism

Trivers, R. L. 1971. The Evolution of Reciprocal Altruism. *Quart. Rev. Biol.,* **46**, 35–57.

Altruism in Man

Sahlins, M. 1976. (*See* his earlier citation under "Kin selection.")

Social Groups

Wilson, E. O. 1971. *The Insect Societies.* Cambridge, Massachusetts, Harvard Univ. Press.

Wilson, E. O. 1975. *Sociobiology. The New Synthesis.* Cambridge, Massachusetts, Harvard Univ. Press.

Wilson, E. O. 1978. *On Human Nature.* Cambridge, Massachusetts, Harvard Univ. Press.

Oster, G. F., and E. O. Wilson. 1978. *Caste and ecology in the Social Insects.* Princeton, New Jersey, Princeton Univ. Press.

Sets of Species

Smith, D. C. 1973. *The Lichen Symbiosis.* Oxford Biology Readers, No. 42. Oxford, England, Oxford Univ. Press.

Richards, A. J. (ed.). 1978. *The Pollination of Flowers by Insects.* London, England, and New York, Academic Press for Linnean Soc. London.

Formation and Evolution of a Faunule

MacArthur, R. H., and E. O. Wilson. 1967. *The Theory of Island Biogeography*. Princeton, New Jersey, Princeton Univ. Press.

Darlington, P. J., Jr. 1943. Carabidae of Mountains and Islands. *Ecological Monographs*, **13**, 37–61.

Darwin Equilibriums

Darlington, P. J., Jr. 1972. Competition, Competitive Repulsion, and Coexistence. *Proc. Nat. Acad. Sci. USA*, **69**, 3151–3155.

Area, Climate, and Evolution

Darlington, P. J., Jr. 1959. Area, Climate, and Evolution. *Evolution*, **13**, 488–510.

Darlington, P. J., Jr. 1957. *Zoogeography. The Geographical Distribution of Animals*. New York, Wiley; reprinted 1980, Krieger.

Darlington, P. J., Jr. 1971. Interconnected Patterns of Biogeography and Evolution. *Proc. Nat. Acad. Sci. USA*, **68**, 1254–1258.

Darlington, P. J., Jr. 1959. Darwin and Zoogeography. *Proc. Am. Philos. Soc.*, **103**, 307–319.

Dobzhansky, T. 1950. Evolution in the Tropics. *American Scientist*, **38**, 209–221.

PART THREE

The Actual Course of Organic Evolution

As buds give rise by growth to fresh buds, and these, if vigorous, branch out and overtop on all sides many a feebler branch, so by generation I believe it has been with the great Tree of Life, which fills with its dead and broken branches the crust of the earth, and covers the surface with its everbranching and beautiful ramifications.

Charles Darwin, *The Origin of Species,* 1859 (quoted from the facsimile of the first edition, page 130).

8

EVOLUTIONARY TRACKS
AND TIMETABLES

Successive levels or grades in the evolution of organisms, from subvital molecules to multicellular eucaryotes, are summarized in Chapter 5. The present chapter is concerned with the phylogenies of separate groups: with the tracks they have followed—or have made as they evolved—and with the times of their beginnings, divergings, and endings.

To cover this subject in one short chapter requires an extreme simplification of an inconceivably complex reality; the actual complexity of phylogenies will be suggested under "Carabid beetles." And many of the histories outlined are partly hypothetical as well as greatly simplified. Hypotheses and guesses are required to make them coherent wholes.

Definitions

Six words used in this chapter are confused so often that they require definition. *Primitive* (from Latin *primus*) means first, or early in an evolutionary sequence; its opposite is *derivative*, which means derived. *Specialized* means adapted to a special environment or special way of life; its opposite is *unspecialized*. Primitive organisms can be specialized. For example, egg-laying monotremes are primitive mammals but are specialized in adaptive structure and behavior (see later in this chapter). *Lower* and *higher*, as used here, refer only to relative positions in evolutionary sequences; lower forms are lower on their evolutionary trees, but are not necessarily inferior in other ways. Some lower forms coexist with higher ones and are still conspicuously successful in competition with them.

The Fabric of Evolution

The evolution of life on earth is comparable to the weaving of a complex fabric, like Darwin's "great tree of life" but with innumerable
branches and twigs forming an interwoven whole more complex than
any tree. The fabric has woven itself, and is still on the loom. The fabric
as a whole is the intricately self-woven biosphere. Within it, bands
represent different groups of evolving organisms. For example, within
about the last third of the fabric a broad band represents all eucaryotes,
and within this band successively shorter and narrower bands represent
vertebrates, mammals, and primates, each successive band being part of
the preceding one. Within each band, narrow ribbons represent evolving species, some of which continually, but irregularly, multiply by spe-
cies-splittings, while others terminate by extinctions. Within each
species' ribbon, strands represent evolutionary continuities of particular
organs, processes, and behaviors. And these are products of genetic
continuities. Genes do evolve, by mutation followed by selection or
random drift. The genes of derivative organisms are, at least in part, different from those of their primitive ancestors, but have evolved from
the latter by genetic continuities. The main bands and a few outstanding ribbons within this fabric are now to be distinguished and traced.

Classification, Geologic Time, and Fossils

To understand phylogenies and how they are traced, readers should
know something about the classification of living things, about geologic
time and how it is measured, and about the fossil record.

Classification.   Most persons have some idea of the major groups of organisms, at least of viruses and bacteria; algae, higher green plants, and
fungi; protozoa (at least amoebae), mollusks, spiders, insects, and some
other conspicuous groups of invertebrates; fishes, amphibians, reptiles,
birds, and mammals; and some of the more obvious orders of mammals.
However, readers who are not systematists are likely to be seriously out
of date, and are urged to refer to a modern classification, for example,
that summarized by Wilson et al. (1973, pages 640–646).

Geologic Time.   Geologists divide the enormous span of time since life
began into eras and periods (Figure 6). The sequence is clear, shown
conclusively by successions of fossil-bearing strata in appropriate stratified rocks, but absolute dating has been difficult.
    The absolute ages of rocks, and of the fossils in them, can now be

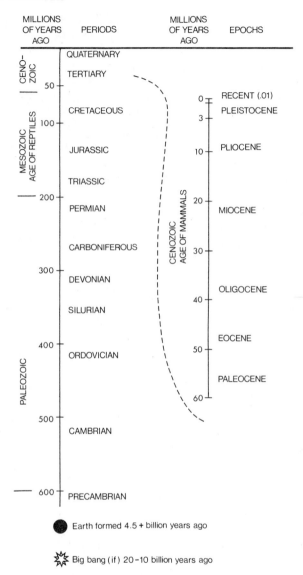

MILLIONS OF YEARS AGO — PERIODS — MILLIONS OF YEARS AGO — EPOCHS

CENOZOIC

QUATERNARY

TERTIARY

50

CRETACEOUS

100

MESOZOIC AGE OF REPTILES

JURASSIC

TRIASSIC

200

PERMIAN

CARBONIFEROUS

300

DEVONIAN

SILURIAN

400

ORDOVICIAN

PALEOZOIC

500

CAMBRIAN

600 PRECAMBRIAN

RECENT (.01)

0

PLEISTOCENE

3

PLIOCENE

10

20

MIOCENE

CENOZOIC AGE OF MAMMALS

30

OLIGOCENE

40

EOCENE

50

PALEOCENE

60

Earth formed 4.5 + billion years ago

Big bang (if) 20–10 billion years ago

Figure 6. Sequence of geologic periods and Tertiary epochs. Boundaries are not shown because they are in fact transitions rather than precisely determined lines, and they have not yet been precisely dated.

determined, within limits, by measuring the decay of radioactive elements, as shown by isotope ratios. Carbon-14 decays (i.e., changes to other isotopes) at a rate that makes it a useful measure of time within about the last 50,000 or possibly 100,000 years. Strontium isotopes change more slowly and give approximate ages of rocks 2 or 3 billion years old. Other elements provide intermediate time scales. And some other methods of dating are used in some cases. Methods of dating are continually being improved, and new methods sometimes force substantial revisions of timetables. An example is the recent redating of the beginning of the Pleistocene from about 1 to 2 or 3 million years ago, which doubled or tripled the time available for the final stages of human evolution. For further details see Harper (1973), especially the table comparing determinations of time scales by 3 methods (page 83), and the discussion of the dating of the Pleistocene (page 357).

The Fossil Record.   The fossil record is literally fragmentary (Chapter 2) —infinitesimal fragments of the quantity and diversity of life of the past —but the fragments are not entirely random. In general, the record is better for organisms with hard parts than with soft parts, and best if the hard parts are informative, as vertebrate skulls are; better for recent times than for ancient times; and better in some places than in others.

Fossils are formed more often in temperate climates than in the tropics, where decomposition is usually more rapid, and they have been looked for most often in the north-temperate zone, where paleontologists are most numerous. The record is therefore biased toward the north in a way that can mislead evolutionists—that, for example, misled Matthew (1915) into postulating northern rather than tropical origins for most groups of animals. Fossilization occurs more often in lowlands, where deposition occurs, than in highlands, where erosion is the rule; this may explain why we have no clear record of the early evolution of angiosperms, if they began as highland plants. And fossils are much more likely to represent points on "fans" than to form linear sequences (Chapter 6C, Figure 5).

Nevertheless, in spite of being fragmentary and biased, the fossil record gives us a surprisingly good view—almost a magical one—of the course of evolution at least of higher plants and animals. What can be done with it is exemplified by Banks' (1970) *Evolution and Plants of the Past* and Romer's (1968) *The Procession of Life*. Readers of Romer's book should remember that, behind the readable text, is a lifetime's accumulation of firsthand knowledge of details by one of the most hardworking, best-informed, and discriminating paleontologists of his generation. A later, much more technical book edited by Hallam (1977) describes the fossil record, patterns of evolution suggested by it, and alternative classifications, but this book is for professionals rather than naturalists.

Kingdoms

The main, partly hypothetical, heavily pruned pattern of phylogenies of living things is diagrammed on a background of the five now-recognized kingdoms in Figure 8 at the end of this chapter. The five kingdoms are:

Procaryotes (bacteria and blue-green algae).

Single-cell eucaryotes.

Green plants.

Fungi.

Animals.

Viruses are excluded from this classification. Their origin is doubtful, and some persons do not consider them living things.

Subvital Molecules; Viruses

As soon as subvital molecules existed capable of reproduction, variation, selection, and evolution—more than 3 billion years ago—phylogenetic diversification presumably began. Molecules on different lineages presumably became adapted to different degrees of salinity, different temperatures, different ways of obtaining and using energy, and different roles in the primitive subbiosphere. Some were, or soon became, photosynthesizers, but others may have been molecular predators or parasites. And although molecular lineages diversified, different lineages may continually have recombined or exchanged parts, so that early phylogenies were partly reticulate.

The phylogenies of viruses are unknown and perhaps unknowable. Their classification ought to reflect their phylogeny, but specialists disagree about virus classification "with the intensity of a religious war." However, the viruses that consist of DNA and those that consist of RNA have perhaps had separate phylogenies.

For naturalists, the best introduction to viruses is probably Williams and Fisher's (1974) *Micrographic Atlas,* which contains fine electron-microscope photographs of 31 viruses, with a few pages of text. Additional reading and references can be found in Carlile and Skehel (1974), especially in Joklik's article "Evolution of Viruses" (pages 293–320). And Fenner (1976) offers a detailed classification.

Bacteria, Blue-Green Algae, and the First Eucaryotes

Apparent bacteria and blue-green algae appear very early in the fossil record and presumably diversified and rediversified, with phylogenies being still partly reticulate as well as linear and diverging. We have no record of the earlier diversifications, but later phylogenetic lines presumably correspond to the principal existing groups, which are described in good textbooks of biology. One or more of these "procaryotes"evolved into the eucaryote cell about 1 billion years ago, by routes that are unknown but are speculated about in Chapter 5, and the first eucaryotes then diversified, different ones becoming ancestral to green plants, fungi and animals.

Ragan and Chapman (1978) review hypothetical phylogenies suggested by earlier authors and propose a biochemical phylogeny of "protists," summarized in their Figure 23. Details are beyond the reach or needs of naturalists.

Green Plants

A "provisional phylogeny" of chlorophyll-bearing organisms is diagrammed by Cloud et al. (1975, page 147). The diagram suggests the origin of all green plants from blue-green algae, the main (complex) diversification of primitive green eucaryotes, and the origin of higher green plants (tracheophytes). The diversification and evolution of the latter are well described in a small, readable book by Banks (1970) and need only be summarized here.

Algae. Apparent eucaryote green algae are fossil in rocks about 1 billion years old, and the further evolution of algae in the Cambrian, Ordovician, Silurian, and Devonian is summarized by Banks (pages 42–50). During this time green algae especially, but also reds and browns, diversified, with selective extinctions; a "general disappearance of the older types of algae and a rise of new ones . . . " apparently occurred during the Devonian (Banks, page 44). Algae, of course, are still dominant in the sea, and some occur in fresh water.

Land Plants. The first green plants on land were probably procaryote blue-green algae on damp surfaces. The first eucaryote land plants probably evolved from fresh-water green algae late in the Silurian and diversified early in the Devonian (Banks, Chapter 4). Some of the early groups soon disappeared. Of more successful, familiar groups, mosses date from the Devonian, or before. Lycopods (clubmosses, and so on) and horsetails appeared and diversified in the Devonian; treelike forms of both groups dominated Carboniferous forests, but disappeared in the Permian; herbaceous forms of both are still common. Ferns appeared early and conifers late in the Carboniferous; cycads and ginkgoes, in the Permian.

Angiosperms. The last plants to become dominant were the angiosperms, or flowering plants, which include most of our familiar green plants except mosses, ferns, conifers and a few others. Although they are numerous, fossilize well, and can be recognized even from fossil pollen, the origin of the angiosperms is still mysterious. The best guess probably is that they are a monophyletic group that evolved from a gymnosperm, perhaps a cycad, somewhere in the tropics early in the Cretaceous, roughly 100 million years ago. Details and alternatives are presented briefly by Sporne (1971) and at greater length in a collection of papers edited by Beck (1976).

Whatever their origin, angiosperms diversified explosively in the Cretaceous, splitting first into "monocots" (grasses, palms, lillies, orchids, and so on) and "dicots" (most other familiar woody plants and herbs). Sporne (in Beck 1976, pages 325–327) offers a hypothetical diagram of their splitting and "circular schemes" of their further phylogenetic diversification.

Early angiosperms were probably woody perennials. Herbaceous forms proliferated late in the Tertiary, especially in the north; their alternation of plants-in-summer and seeds-in-winter was evidently an adaptation to increasingly seasonal northern climates.

This too-brief sketch of the history of green plants should at least give naturalists the beginning of an understanding of the weaving—self-weaving—of the green-plant fabric that now covers the earth.

Fungi

Fungi are not degenerate green plants but an independent kingdom with a separate origin from one or more single-cell eucaryotes (Figure 8). Fungi have few structural characters; they do not fossilize well; their classification and phylogeny are therefore difficult; and the names of the different groups are formidable—"Heterobasidiomycetes," "Hyphochytridiomycetes," and so on—and unknown to most of us. They will therefore be summarily dismissed here. Readers who wish to pursue the difficult subject of their phylogeny can begin with Klein and Cronquist's (1967) paper, with a diagram of "alternate proposed phylogenies of the fungi" on page 255 and a good reference list.

Lichens

Lichens are of great interest to naturalists for the roles they play especially in inhospitable environments including the High Arctic, and for their dual personalities: they are symbiotic sets-of-two, each consisting of a supporting fungus and a photosynthesizing alga. Different groups of fungi and algae are represented in different lichens. The latter are therefore convergent symbionts which have no independent phylogeny. Smith (1973) provides a good introduction to them.

Animals: Protozoa

The first animals were single-celled eucaryote protozoa of which the descendants now include amoebas, cilates, and flagellates (from which higher, multicellular animals are probably derived), parasites (including the blood parasites that cause malaria), and other less familiar forms. Their ancestors were one or more single-celled eucaryote algae, but details are in doubt, and so is their later phylogeny. Readers who wish to pursue this also-difficult subject can go to Hanson (1977), especially his Chapter 15 (phylogenetic conclusions), but must be prepared to learn new phylogenetic terms and new procedures as well as many unfamiliar group names.

Multicellular Invertebrates

The first multicellular animals were evidently derived from one or more ciliate or flagellate protozoa. Romer (1968) gives an outstandingly well-informed, readable account of the phylogenies of the principal groups of multicellular animals from these small beginnings. He brings the animals as near to life as a conscientious paleontologist could, and much better than I can do in smaller space. I shall therefore treat here, briefly, chiefly the more important groups with living representatives, but shall give more details in two special cases: that of the arthropods (stressing beetles and Carabidae) in the present chapter and that of mammals ancestral to man in Chapter 9. All the principal phyla of multicellular animals have now been found fossil in the Cambrian.

Sponges. The simplest true multicellular animals are the sponges (Porozoa), which, as Romer says, are comparable to colonies of single-celled flagellates in which different individuals are differently specialized and interdependent. They were fairly diverse early in the Cambrian, and they still exist, most in the sea but a few in fresh water. However, they were apparently an evolutionary sideline, not ancestral to other multicellular animals.

Coelenterates. The main line of evolution of multicellular higher animals (metazoa) may have begun with the origin of coelenterates from a ciliate or flagellate protozoan. Coelenterates began perhaps as relatively simple forms comparable to existing fresh-water hydras, then diversified into (among other groups) jellyfishes and corals. Corals, or structures formed by them, are varied as far back as the Devonian, but impressions that may have been made by jellyfishlike coelenterates occur in much older rocks, even before the Cambrian, and "may have been close to the truly ancestral metazoan types" (Romer, page 43).

Worms. Worms have probably radiated from ancestors comparable to existing planarians, but "worms" are diverse and not all closely related among themselves, and their phylogenies are poorly known. Annelids include segmented worms with a row of leglike appendages on each side. Earthworms and leeches are probably specialized derivatives of (probably originally marine) annelids. And mollusks and arthropods too are probably derived from annelids.

Mollusks. Mollusks are probably derived from annelid worms, although they have lost the annelids' segmentation and appendages. The ancestral mollusk probably had a simple shell over a still-segmented body; the shells, given the name *Pilina*, are fossil in the older Paleozoic; the animals were thought to have been long extinct until, not many years ago, living ones (named *Neopilina*) were dredged from deep ocean bottoms.

From such an ancestor, mollusks radiated an astonishing diversity of differently adapted groups, including chitons, gastropods (snails), pelecypods (clams), and cephalopods (squid, octopus, nautilus, and so on). The cephalopods especially, although few in number now, had a complex phylogeny, with apparently three successive major radiations.

Arthropods, Jointed-Legged Invertebrates

Arthropods are armored, segmented, jointed-legged invertebrates (with other less obvious distinguishing characteristics). They probably evolved from centipedelike annelid worms early in, or before, the Cambrian. Modern specialists (Anderson, 1973; Manton, 1977) recognize three or four phylogenetically separate groups of them.

One group includes the extinct marine trilobites (Cambrian to Permian); the marine Chelicerata (Cambrian to Present), most extinct, but including the horseshoe crabs (Xiphosura) of which a few still exist along the coasts of northeastern North America and southeastern Asia; and the arachnids (Silurian to Present), including existing terrestrial scorpions, spiders, and their relatives, which made the first major arthropod invasion of the land. The second group includes the mostly marine and fresh-water crustaceans (Cambrian to Present), including shrimps, crabs and lobsters, barnacles, and so on. And the third group is primarily terrestrial and is in turn phylogenetically tripartite. It includes the onychophorans (Cambrian or before to Present), of which "peripatus" still exists in the tropics and southern hemisphere; the myriopods (Upper Silurian to Present), including millepedes and centipedes; and the hexapods, chiefly insects (probably late Devonian to Present). All these groups evolved complexly, but only the hexapods can be considered in more detail here.

Insects. The Hexapoda—six-legged arthropods, mostly insects—was the second major arthropod group that dominated the land. The ancestor of the group was probably centipedelike (but not a centipede), and its evolution involved a succession of major adaptive levels and radiations and a bewildering number of phylogenetic lines. The enormous numbers, diversity, and success of insects are commonplaces of everyday life. But their sphere has limits. They do dominate the land and they are abundant also in fresh water, but they are insignificant in the sea. A few specialized insects live on beaches between the tide lines, in tide pools, or in shallow sea water; and one small group of water striders lives on the surface of the open ocean, but otherwise insects have somehow been barred from the huge, rich marine environment. No ecologic limitation of any organisms is more striking than this!

Entomologists recognize four principal, successive levels of evolution and radiation of insects.

First was a (probably moderate) diversification of wingless, chiefly terrestrial forms, of which a few (silverfish, and springtails if the latter are considered insects) still exist.

Second, one of the primatively wingless insects evolved wings. We do not know what or where it was, but it had probably entered fresh water. The most primitive existing winged insects develop (spend most of their lives) in fresh water; wings might have evolved from swimming or stabilizing flanges; and the first use of flight may have been (as it still is for many insects) dispersal in discontinuous, fluctuating habitats such as fresh-water ones. The first winged insects, classified as Paleoptera, could move the wings only up and down. They were fairly diverse by the Carboniferous, and some—mayflies, dragonflies, and damselflies—still exist. They were the first flying animals of any sort, and presumably dominated the air, where they had no preditors or competitors to face except each other. But they could not fold their wings backward out of the way, so that the adults were awkward on the ground and unable to enter the almost endless microhabitats that higher winged insects occupied later. They were (and their descendants are) comparable to airplanes which, because of their wide-spreading wings, cannot run along narrow roads or through woodland.

A third radiation followed the evolution of a hinge that allowed the wings to be folded backward out of the way. This did allow diverse adaptations to terrestrial habitats. The first insects that could do this had a relatively simple pattern of development, the immature stages being in general like the adults; the adults alone could fly, but the wings developed externally, as visible flaps or pads in the immatures (called nymphs). Some of these insects were dramatically successful, and still are. Among them are the cockroaches, termites, grasshoppers, true bugs, and their allies.

The fourth radiation of insects followed evolution of a new pattern of ontogeny, in which the wings develop internally in the immatures (now called larvae), are reversed in a transition stage (the pupa), and are expanded in the adults. This allows extreme differentiation of larvae and adults not only in anatomy but also in way of life; one individual insect can, first as larva and then as adult, live in two quite different places, eat entirely different foods, and have entirely different adaptations and specializations. Such insects can make the best of two worlds, and are outstandingly successful. They include the lacewings, true flies, bees and wasps, butterflies and moths, and beetles.

Beetles. Each principal insect order has had its own complex phylogeny following an initial evolutionary innovation. The beetles—order Coleoptera—are a conspicuous example. Beetles originated probably from a megalopteran (something like a lacewing) perhaps in the Permian. Their innovation was the modification of the forewings into additional adult body-armor which gave greatly increased protection—against desiccation, diseases, and environmental chemicals as well as predators—without sacrificing the power of flight. Most beetles can still fly; they have wings that remain folded under their wing cases until needed. The beetles have therefore retained and added together the evolutionary advantages gained in the four principal stages of insect evolution—terrestrial life, power of flight, wing folding, and differentiation of larva and adult

—and added a fifth innovation of their own—more effective adult armor. The result has been a spectacular radiation, suggested by Evans (1975; but readers of this book should read also Hinton's somewhat critical review of it cited in my chapter reference list).

Carabid Beetles. Carabids exemplify the numbers and diversity of beetles and the complexity of phylogenies in general. They are briefly described in Chapter 3. Now, something more is to be said about their evolution. Their ancestor and the place and time of their origin are all unknown; no significant fossil carabids have been found, except a few in Baltic amber. But their present diversity surely reflects a complex phylogeny, with many different lines evolving and spreading over the world, many still successful, but with some primitive relicts scattered among them. The pattern suggests partly successive and partly contemporaneous radiations of many separate tribes and genera, with much extinction as well as many survivals, so that the existing carabid fauna is the product of a complex "cumulative sequence."

This summary does not sufficiently emphasize the complexity of evolution of this family of beetles. Each of the many phylogenetic lines within the family has had its own, probably complex history, which has probably often involved several successive radiations. In most cases the details are wholly unknown, but the evolutionary history of two special groups of carabids can at least be guessed about.

Tiger Beetles. The tiger beetles (subfamily Cicindelinae) are a well-defined group of active predators of which the apparent phylogeny and present distributions suggest three successive major radiations: first of a primitive arboreal stock, which now survives only in Madagascar (not Africa) and tropical America; then of several terrestrial stocks, which are now widely but discontinuously scattered over the world; and finally of a very successful terrestrial stock, the genus *Cicindela* in a broad sense, which has dispersed apparently from Africa recently and complexly over all the habitable continents. One reason for its success may be the evolution of a dense mat of white scales covering the lower surface, which insulates the insects from the hot ground on which the adults of many species live. (This history is from the unpublished manuscript cited in the chapter reference list.)

Paussids. The other carabid group of which the evolutionary history can be reasonably guessed about is the tribe Paussini, the paussids, most of which live in ants' nests and all of which are probably associated with ants in some way. They are apparently derived from a primitive (ozaenine) carabid ancestor, and their relatively rapid evolution (see Chapter 6B) has apparently involved a succession of events diagrammed in Figure 7, including perhaps nine distinguishable radiations.

These are only examples of the probable complexity of evolution of

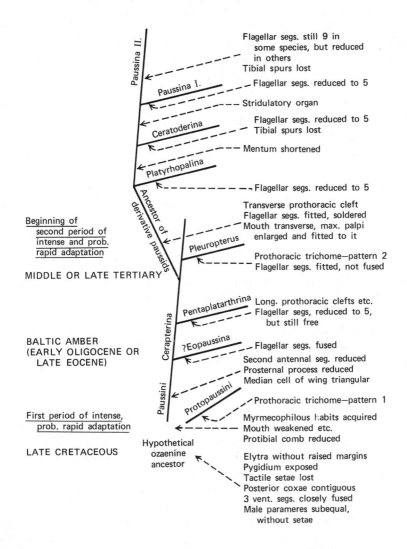

Figure 7. Diagram of selected, significant events in the evolution of tribes and subtribes of paussids (from P. J. Darlington, Jr., 1950, Paussid Beetles, *Trans. Am. Entomol. Soc.* lxxvi, p. 82). Note the amount of evolution that has occurred since the formation of the Baltic amber, while (some) ants have evolved only slightly differentiated species.

many other tribes within the family carabidae. To get some idea of the phylogenetic complexity of the family as a whole, these examples should probably be multiplied by at least a hundred to cover adaptive radiations of extinct and living tribes, and this total should then be multiplied again to cover evolutions of thousands of genera and tens of thousands of known and still-unknown species. And this total must be multiplied by at least ten for the Coleoptera as a whole, with comparable additions for all the other insects. This is the kind of complexity within which we try to trace phylogenies and reduce them to simplicity!

Echinoderms

The last few pages have summarized a major band in the fabric of evolution leading from coelentrates to worms to (in one direction) mollusks and (in another) to arthropods, insects, beetles, and Carabidae. Another band apparently runs from or through echinoderms to vertebrates.

Romer (1968, pages 116ff) prefaces his account of echinoderms with a brief discussion of "filter feeders," including bryozoans (moss animalcules) and brachiopods (lamp shells) as well as primitive echinoderms. Ancestral echinoderms probably *were* sessile filter feeders. A considerable variety of "echinoderm types of archaic nature," all belonging to groups now long extinct, are fossil in the Cambrian (Romer, page 126). Later, separate radiations occurred of crinoids (sea lilies) which still survive in the sea in fairly deep water, and of starfishes and their allies, sea urchins, and sea cucumbers.

Chordates to Vertebrates

The origin of vertebrates is doubtful and may have been somewhat devious. Romer (1968, pages 131ff) summarizes some earlier ideas about it. His theory is that the line began with a sessile, filter-feeding animal comparable to a primitive echinoderm. A change then occurred in method of filter feeding, from "arm gathering" to internal filtering, producing a still sedentary animal with gill slits, comparable to existing tunicates; it had a free-swimming larva (which many tunicates still have), with gill slits, a simple vertebratelike nervous system and sense organs, and a notochord. And then the sessile adult was suppressed— this being an example of neoteny—and the larva became the adult and evolved to chordate and vertebrate levels.

Fishes. The first vertebrates were fishlike, jawless filter feeders, which are now known to have originated before the end of the Cambrian (Repetski, 1978), which radiated in the mid-Paleozoic, then disappeared from the fossil record; but a specialized group of them still survives, the eel-like, parasitic lampreys, which still have filter-feeding larvae. Later fishes went through a series of evolutionary stages, radiating or reradiating at each stage, for more than 100 million years before the beginning of land-living vertebrates. Cartilaginous fishes (sharks, and so on) split from the bony fishes in the Devonian and have had their own complex phylogeny. Higher bony fishes had appeared early in the Devonian (if not before) and split into ray-finned and fleshy-finned groups, and while the former diversified spectacularly—most existing fishes are ray-fins or their derivatives—the more modest radiation of the fleshy-finned fishes produced the lobefin that moved from water to land and evolved into the first amphibians, in the Devonian.

Amphibians. On land, amphibians diversified in the Carboniferous and Permian, then declined. Salamanders, frogs, and caecilians are the three surviving, specialized groups.

Reptiles. The rise of reptiles followed the evolution of the "amniote egg," "the most marvelous single invention in the whole history of vertebrate life" (Romer, 1968, page 183), which could develop on dry land and which freed the amphibian ancestor of reptiles from dependence on water. (Many other amphibians have "invented" other kinds of eggs which allow them to omit an actual aquatic larval stage but still require a damp environment; for example, the small *Eleutherodactylus* frogs that naturalists hear calling in the rain forest in Puerto Rico lay their eggs on land in damp places; the embryos are somewhat tadpolelike, but the eggs hatch directly into tiny frogs.)
    The first reptiles evolved from anthracosaur amphibians (which themselves soon became extinct) late in the Paleozoic, and primitive reptiles diversified in the Carboniferous and Permian; the reptile ancestors of mammals can be traced from this time, evolving inconspicuously among the "ruling" reptiles. Later, in the Mesozoic—the Age of Reptiles—three principal groups diverged and radiated. (1) Turtles have continued "relatively uneventfully" since then (Romer). (2) Lepidosaurs produced rhynchocephalians, of which the Tuatara on New Zealand is the sole survivor; lizards; and snakes, which evolved from a *Varanus*-like lizard. And (3) archosaurs diversified into crocodiles and so on, dinosaurs, flying pterodactyls, and birds. (The gigantic marine plesiosaurs and ichthyosaurs had separate origins.)

The diversity of reptiles is barely suggested by this sketch of their evolution. Romer (1968) describes them in more detail, defines the unfamiliar names used in the preceding paragraph, and summarizes reptile radiations in his Figure 44. Only a few comments are needed here. New England naturalists should know that lizards are far more numerous than snakes almost everywhere; northeastern North America is the only part of the world where snakes are numerous and lizards absent; and it is also almost the only continental area where we can safely teach our children not to fear snakes. The Komodo Dragon, the largest living lizard, confined to a few small islands in the Malay Archipelago, is sometimes referred to as if it were a unique relict—a living fossil—but it is not; it is only the largest of living monitor lizards—*Varanus*—of which other species are common throughout the warmer parts of Asia, Africa, and Australia. The great interest of dinosaurs, to readers of all ages, is reflected in a spate of recent books, of which only one is listed at the end of this chapter; the extinction of dinosaurs is considered in Chapter 6C. And the evolution of the reptile that became a bird is noted next.

Birds.   Birds evolved in the Jurassic from a reptile, not from a flying pterosaur but from a small dinosaurlike ancestor. Later, birds radiated and reradiated, toothed birds in the Cretaceous, toothless prepasserines, beginning perhaps later in the Cretaceous (but details are unknown), and finally and explosively passerines (perching birds, including song birds) mainly in the Tertiary (but again details are mostly unknown). The last two radiations of toothless prepasserines and passerines, produced the existing diversity of birds.

Jurassic *Archaeopteryx*, although classified as a bird (because it had feathers and for other reasons) and considered ancestral to the birds, was still reptilelike in many ways. Whether or not it could fly has been debated recently, but it probably could. Its structure apparently does not forbid "powered flight" (Olson and Feduccia, 1979); its wing feathers were unsymmetric, like those of flying birds (Feduccia and Tordoff, 1979); and its brain was larger than the brain of a reptile of the same size, which suggests that it had begun to evolve the complex neurological mechanism of flight. Imaginative naturalists may therefore reasonably visualize *Archaeopteryx* flapping up into a Jurassic tree fern to escape a predaceous dinosaur.

Mammals: Three (Four?) Radiations.   Mammals originated from therapsid reptiles perhaps early in the Jurassic, evolved inconspicuously among the ruling reptiles for 100 million years, and then radiated spectacularly, their rise to dominance perhaps following evolution of effective temperature control, improved reproduction, and a better brain.

During the Tertiary—the Age of Mammals—three mammal faunas radiated in different parts of the world. They illustrate the power of adaptive evolution to fill niches and to produce harmonic faunas, with much parallelism and convergence, and naturalists need to understand their history in order to understand what they see now in different regions.

1. In the main part of the world—Eurasia, Africa, and North America —a diversity of placentals evolved from insectivorelike ancestors. Many, including some bizarre giants, became extinct from time to time and were replaced by presumably competing groups. Products of this radiation now include still-primitive insectivores; primitive and derivative primates; placental carnivores; a diversity of hoofed mammals; elephants and hyraxes; the Aardvark and Pangolin; rodents; rabbits, which are not rodents but have evolved separately; bats in the air; and three groups that have entered the sea independently: whales and porpoises, seals, and manatees (and, also separately, the Sea Otter and amphibious Polar Bear). This radiation has, of course, been infinitely more complex than can be described here, with subradiations in Africa, Eurasia, and North America, and complex mixings, extinctions, and replacements.

2. Australia was an island continent during the Tertiary, as now, and marsupials radiated separately there, many of them converging with placentals. Different Australian marsupials resemble shrews or weasels or wolves, rabbits, or squirrels, and the latter include plant eaters, insect eaters, and three separate stocks of flying squirrel-like gliders. There are also the well-known Koala, an arboreal leaf eater; wombats, which resemble giant, ground-living rodents; a marsupial anteater; an extraordinary marsupial mole; and kangaroos, which crop grass and are ecological equivalents of placental hoofed grazers elsewhere. There are no marsupial equivalents of small, gnawing rodents (the so-called marsupial rats and mice do not have gnawing teeth) or of bats, but true placental rats and bats reached Australia from Asia during the Tertiary. And there are no aquatic marsupials in Australia, which perhaps accounts for the survival of the Platypus.

3. South America too was an island continent during most of the Tertiary, and a separate radiation of mammals occurred there, derived from a very limited set of ancestors which included both marsupials and placentals. The products of this radiation included various opossums; marsupial carnivores, all now extinct, one of them a counterpart of a saber-toothed tiger; an almost unbelievable diversity of hoofed placentals of peculiar orders, all extinct before the coming of man, and unknown

even by name to most naturalists; edentates, including giant sloths and giant armadillolike forms; monkeys; rodents of peculiar groups; and bats. And when South and North America were connected in the Pliocene, a complex exchange of mammals occurred, followed by selective extinctions, the result being the replacement of much of the old South American mammal fauna by invaders from the main part of the world, including placental carnivores, modern ungulates, rodents, and some others, so that the present South American fauna is a mixture of parts of the old South American mammal fauna and newcomers. In the other direction, only one mammal of South American origin now occurs far north in North America—our Porcupine.

4. To return to Australia, monotremes *may* have radiated there before marsupials. Surviving monotremes are primitive but specialized (see "Definitions" earlier in this chapter); they still lay eggs, but are otherwise more specialized than any marsupials, the aquatic Platypus to grubbing ducklike for food under water, the Echnida to breaking into termite nests and feeding on termites. Monotremes are not known ever to have occurred outside the Australian Region, and their fossil record even in Australia is insignificant. However, the extreme specialization of the survivors suggests that they are at the ends of lines of adaptive evolution longer than those of any marsupials, and this in turn suggests that monotremes radiated in Australia before marsupials did, that they may once have been as diverse as marsupials are now, and that marsupials have replaced all of them except the two that are so specialized that they are almost beyond the reach of competition. This is speculation, but it fits what few pertinent facts there are—and the extraordinarily specialized Platypus and Echnida must have evolutionary histories. Both, incidentally, range widely across climatic zones, from cool-temperate Tasmania to tropical North Queensland (and the Echnida has close relatives in New Guinea), and both are still common in some places, so that visiting naturalists can expect to see them.

An old idea held that marsupials were older than placentals, and that Australia was occupied by marsupials and then cut off from the other continents before placentals evolved, but this idea was wrong. Marsupials and placentals are parallel, not successive, groups; their separate ancestors are known fossil well back into the Mesozoic. Why marsupials reached Australia first is debated; Keast et al. (1972, pages 70–72) review recent theories. I think that one or more marsupial ancestors probably reached Australia across an early Tertiary water barrier that placentals did not cross—until, later in the Tertiary, *several* rodents reached Australia from Asia across water barriers that other mammals did not cross (excepting bats).

In any case, the radiations of placental mammals in the main part of the world, marsupials in Australia, and both in South America occurred in different places, not at different times. Only the monotreme radiation (if there was one) was earlier.

The complex geographic evolutionary history of mammal faunas is summarized in more detail in my *Zoogeography* (1957). Romer's *Procession* (1968) makes more of a narrative especially of the radiation of placentals in the main part of the world. Australian marsupials are described by Tyndale-Biscoe (1973), and their biology in more detail by Stonehouse and Gilmore (1977). Keast et al. (1972) treat the evolution of mammals on the southern continents in an important book which naturalists can at least sample profitably. And the history of the line of placental mammals that led to man is continued in Chapter 9.

Summary

The main-line tracks and timetables of organic evolution are summarized in a heavily pruned phylogenetic diagram in Figure 8, which shows only major levels and origins, traces to the present only a few groups especially discussed in this book, and leaves out most other complicating details.

READING AND REFERENCES

Classification, Geologic Time, and Fossils

Classification

Wilson, E. O., et al. 1973. *Life on Earth.* Stamford, Connecticut, Sinauer Associates.

Geologic Time

Harper, C. T. (ed.). 1973. *Geochronology. Radiometric Dating of Rocks and Minerals.* Stroudsburg, Pennsylvania, Dowden, Hutchinson & Ross.

The Fossil Record

Matthew, W. D. 1915. Climate and Evolution. *Ann. New York Acad. Sci.,* **24,** 171–318. Reprinted 1939 as *Special Publ. New York Acad. Sci., No. 1.*

Banks, H. P. 1970. *Evolution and Plants of the Past.* Belmont, California, Wadsworth Publ. Co.

Romer, A. S. 1968. *The Procession of Life.* London, England, Weidenfeld and Nicolson; Cleveland, Ohio, and New York, World Publ. Co.

Hallam, A. (ed.). 1977. *Patterns of Evolution as Illustrated by the Fossil Record.* New York, Elsevier. (A galaxy of important but technical papers, usefully reviewed by Simpson in *Nature* **273,** 77–78, and by Ricklefs in *Science* **199,** 58–60.)

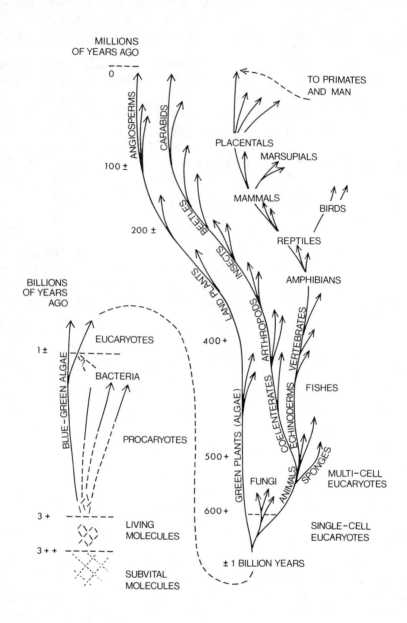

Figure 8. Phylogenetic diagram of the evolution of life on earth, simplified and heavily pruned, showing only major origins, and following to the present only three groups of organisms specially considered in the text.

Subvital Molecules; Viruses

Williams, R. C., and H. W. Fisher. 1974. *An Electron Micrographic Atlas of Viruses.* Springfield, Illinois, Charles C. Thomas.

Carlile, M. J., and J. J. Skehel. 1974. *Evolution in the Microbial World.* Cambridge, England, Cambridge Univ Press.

Fenner, F. 1976. *Classification and Nomenclature of Viruses.* Second Report Intl. Comm. Taxonomy of Viruses. Basel, S. Karger.

Bacteria, Blue-Green Algae, and First Eucaryotes

Ragan, M. A., and D. J. Chapman. 1978. *A Biochemical Phylogeny of the Protists.* New York, Academic. [These authors (page 2), after discussing other usages, define "protists" broadly as including procaryotes as well as simple eucaryotes. Some other authors prefer to restrict the term to the latter.]

Green Plants

Cloud, P., M. Moorman, and D. Pierce. 1975. Sporulation and Ultrastructure in a Late Proterozoic Cyanophyte: Some Implications for Taxonomy and Plant Phylogeny. *Quart. Rev. Biol.,* **50**, 131–150.

Banks, H. P. 1970. Cited previously under "The fossil record."

Angiosperms

Sporne, K. R. 1971. *The Mysterious Origin of Flowering Plants.* Oxford Biology Readers, No. 3. Oxford, England, Oxford Univ. Press.

Beck, C. B. (ed.). 1976. *Origin and Early Evolution of Angiosperms.* New York and London, England, Columbia Univ. Press.

Fungi

Klein, R. M., and A. Cronquist. 1967. A Consideration of the Evolutionary and Taxonomic Significance of Some Biochemical, Micromorphological, and Physiological Characters in the Thallophytes. *Quart. Rev. Biol.,* **42**, 105–296.

Lichens

Smith, D. C. 1973. *The Lichen Symbiosis,* Oxford Biology Readers, No. 42. Oxford, England, Oxford Univ. Press.

Animals: Protozoa

Hanson, E. D. 1977. *The Origin and Early Evolution of Animals*. Middletown, Connecticut, Wesleyan Univ. Press.

Multicellular Invertebrates

Romer, A. S. 1968. *The Procession of Life*. (Cited earlier under "The fossil record.")

Arthropods, Jointed-Legged Invertebrates

Anderson, D. T. 1973. *Embryology and Phylogeny in Annelids and Arthropods*. New York, Pergamon.

Manton, S. M. 1977. *The Arthropods: Habits, Functional Morphology, and Evolution*. Oxford, England, Oxford Univ. Press.

Beetles

Evans, G. 1975. *The Life of Beetles*. London, England, George Allen & Unwin. Reviewed by Hinton, 1975, *Nature*, **258**, 554.

Carabid Beetles

See Chapter 3 under "Carabid beetles."

Tiger Beetles

Darlington, P. J., Jr. *The Geography of Tiger Beetles*. Unpublished manuscript.

Paussids

Darlington, P. J., Jr. 1950. Paussid Beetles. *Trans. Am. Entomol. Soc.*, **76**, 47–142.

Echinoderms

Romer, A. S. 1968. *See* earlier citation under "The fossil record."

Chordates to Vertebrates

Romer, A. S. 1968. (*See* earlier citation.)

Romer, A. S. 1966. *Vertebrate Paleontology*. Chicago, Univ. Chicago Press.

Stahl, B. J. 1974. *Vertebrate History: Problems in Evolution*. New York, McGraw-Hill. [This text does not replace Romer's but allows students to get a preliminary understanding "without becoming surfeited by facts" (Barry Cox).]

Løvtrup, S. 1977. *The Phylogeny of Vertebrata.* New York, Wiley. [Ignores the fossil record and proposes a phylogeny based on existing forms arranged according to the rules of "cladism." The book may be important, but it is, as the author says (page 285), "argumentative rather than descriptive," and no one should try to read it who does not already have a good, detailed knowledge of vertebrates. It is reviewed in *Systematic Zoology,* **27**, 244–248, and in *Nature* **269**, 269–270.]

Fishes

Repetski, J. E. 1978. A fish from the Upper Cambrian of North America. *Science,* **200**, 529–530.

Greenwood, P. H. 1975. [J. R. Norman's] *A History of Fishes,* 3rd ed. London, England, Ernest Benn.

Lagler, K. F., J. E. Bardach, R. R. Miller, and D. R. M. Passino. 1977. *Ichthyology,* 2nd ed. New York, Wiley.

Amphibians

Noble, G. K. 1931. *The Biology of the Amphibia.* New York, McGraw-Hill; reprinted 1954, New York, Dover. (Old, but still useful.)

Frazer, J. F. D. 1973. *Amphibians.* Wykeham Sci. Svc., Vol. 25. New York, Springer-Verlag.

Reptiles

Romer, A. S. 1968. *The Procession of Life.* (Cited earlier.)

McLoughlin, J. C. 1979. *Archosauria: a New Look at the Old Dinosaur.* New York, Viking. (A good, short, popular introduction to dinosaurs and their relatives.)

Birds

Thomson, Sir A. L. 1964. *A New Dictionary of Birds.* London, England, Thomas Nelson, for British Ornithologists Union.

Olson, S. L., and A. Feduccia. 1979. Flight Capability and the Pectoral Girdle of *Archaeopteryx. Nature,* **278**, 247–248.

Feduccia, A., and H. B. Tordoff. 1979. Feathers of *Archaeopteryx;* Asymmetric Vanes Indicate Aerodynamic Function. *Science,* **203**, 1021–1022.

Mammals

Darlington, P. J., Jr. 1957. *Zoogeography. The Geographical Distribution of Animals.* New York, Wiley; reprinted 1980; Krieger.

Romer, A. S. 1968. *The Procession of Life.* (Cited earlier.)

Halstead, L. B. 1978. *The Evolution of the Mammals.* London, Peter Lowe. (Parts of the text are glib rather than accurate, and many of the "pictures of over 200 animals in full colour and line" are not of real animals but lurid reconstructions, the "full colour and line" being imaginary. However, naturalists who can allow for these faults will find the book a useful, short introduction to mammal evolution.)

Tyndale-Biscoe, H. 1973 *Life of Marsupials.* London, England, Edward Arnold.

Stonehouse, B., and D. Gilmore (eds.). 1977. *The Biology of Marsupials.* Baltimore, Univ. Park Press.

Keast, A., F. C. Erk, and B. Glass (eds.). 1972. *Evolution, Mammals, and Southern Continents.* Albany, State Univ. of New York Press.

9

HUMAN EVOLUTION

To understand how and why we have evolved into what we are, we must begin " . . . far back in the days that shaped us, in the days none knew, and that very few thought mattered" (Dunsany, *The Blessings of Pan*). To do this should fascinate naturalists who want to understand themselves and their place in the world, and I shall try to write about it as an evolutionist-naturalist writing for other naturalists.

In writing this chapter almost more than in any other I have been overwhelmed by the mass and diversity of known details and my temerity—I hope not effrontery!—in trying to trace a simple path through the maze of details accumulated by specialists.

For convenience of reading and reference this important chapter is divided into four sections. Section A traces the living, evolving "ribbon" that wove itself toward man through the fabric of evolution from the origin of life to the beginning of the Age of Mammals; B, stages of evolution from tree shrew, to monkey, to ape, to ape man: C, from ape man to man; and D, special aspects of evolution of brain and intelligence, and some consequences.

Sources for a History

An evolutionary history of man has to be derived partly from evidence and partly from inference. Evidence includes fossils and, later, other objects preserved from the past (living sites, refuse, primitive tools, cave paintings, and so on). Inferences are derived partly from comparison of man with simpler organisms which, although they are collateral relatives rather than actual ancestors, seem to represent stages in man's evolution, and partly from "extrapolation backward" from man as he now is, reversing the adage to read "As the tree is inclined, so was the twig bent."

SECTION A.   FABRIC AND RIBBON: BEGINNINGS TO MAMMALS

The Fabric of Evolution and the Ribbon to Man

The evolution of life on earth is compared in Chapter 8 to the self-weaving of an intricately complex fabric. In the beginning, during the origin of life and for a while afterward, different evolving "living molecules" and, later, simple organisms may have exchanged parts (genes) in ways that made the fabric reticulate, but after the origin of eucaryotes one narrow species-ribbon within the complex fabric of evolving life has woven itself toward man. Whenever an ancestral species has split into daughter species, one of the latter was (unknowingly) on the way to man. But even the most imaginative and foresighted naturalist, if present then, could not have distinguished the actual ancestral species from its "sisters" and could not have predicted where the evolving species-ribbon would lead. Now, looking back from where we stand at the growing point of the still evolving ribbon, we know that a narrow ribbon no wider than a single species (or perhaps dispersed or divided at first, and later blurred a little by occasional hybridizations) must have woven itself through all the complex (and in detail largely unknown) origins and successive radiations of procaryotes, early eucaryotes, the first multicellular animals, prechordates and chordates, fishes, amphibians, reptiles, early mammals, insectivores, early primates, monkeys, apes, and ape men, to us now. We can say what some of the beginnings, markers, choices, and continuities were that made and marked the ribbon, and we can place the ribbon in particular bands within the fabric as a whole, but we cannot see precisely where the ribbon was most of the time, and (as I have said) a naturalist could not have distinguished it.

Beginnings, Markers, and Continuities

Some markers and genetic continuities in man's evolutionary ribbon trace from the beginning of life: for example, right-handed DNA and the genetic code. Others date from successive early stages of evolution: from precellular life, our ATP-and-oxygen-using metabolism; from the first eucaryote cells, how each cell is organized and the mechanisms of cell division and Mendelian heredity; and from prechordates, the pattern of cell and tissue differentiation that begins our individual ontogenies. And additional markers and continuities in man's evolutionary ribbon date from successive chordate-vertebrate stages: from chordates and early fishes, the beginnings of heart, blood, bone, kidneys, and even lungs; from the fish-amphibian transition, improved lungs and kidneys and five-digit hands and feet; from amphibians and reptiles on land, continual improvements in locomotion, physiology, and reproduction; and from early mammals, hair, warm-bloodedness, placental reproduction, improved brain, and increasingly complex individual and social behavior.

This is not intended to be an erudite recapitulation but only an informal sample of some of the more obvious markers and continuities that have been woven into man's evolutionary ribbon from time to time, and that gradually made a ribbon that finally self-wove itself into man as he now is. It is intended to emphasize that what happened during the origin of life and stage by stage afterward is indeed part of man's evolutionary history.

Under a Magic Spotlight

To bring man's evolutionary ribbon more to life, imagine a magic spotlight focused unerringly on the actual living individuals that carried the genes that, themselves evolving, went into the making of man, and imagine what a naturalist might have seen from time to time.

In the beginning, during and soon after the origin of life, the spotlight might have illuminated, not a single lineage, but a network of lineages of self-reproducing, evolving molecules which continually exchanged parts of themselves. The naturalist, standing on naked, deeply eroded land and looking where the light pointed, would have seen nothing with the unaided eye except (tinted?) water, but with a good microscope he might have seen in the water a variety of almost-living and living molecules, each kind behaving in its own way and competing with other kinds, some using radiant (ultraviolet) energy from the sun to act, grow, and reproduce, and others perhaps already preying on the radiation users, and each (if illuminated by the spotlight) carrying genes that, themselves evolving, joined in the weaving of man's evolutionary ribbon.

Later (if the eucaryote cell originated by a symbiotic coming together of previously separate forms of life; see Chapter 5) the spotlight might have illuminated several different microscopic organisms at the same time. The naturalist, still standing on deeply eroded land and with no plant cover except perhaps a skim of blue-green algae on damp surfaces, still looking into water, and still using a microscope, might have seen something like an amoeba, a food taker preying on other microscopic organisms as amoebas do, but dividing only in a relatively simple way; a blue-green alga, one particular species among many, photosynthesizing its own food; and a spirochetelike bacterium, also a food taker or parasite, dividing in a special, precise way. If the eucaryote cell did indeed originate by a coming together of organisms like these, each separate kind illuminated by the spotlight, weaving its separate evolutionary ribbon, carried genes that (modified) eventually wove themselves into man's ribbon. The naturalist might later have seen (perhaps) the amoeboid ingest the alga and (perhaps) the spirochete attack them both, and might then have seen (perhaps) what began as a prey-predator-parasite association become a symbiotic whole. (If this happened, it was an early example of "forced altruism" evolving into reciprocal altruism; see Chapter 7.)

Later again, but still before dense vegetation covered the land, the naturalist might have seen under the spotlight, still in water but now with the unaided eye, a tunicate, looking like "a small blob of nothing in particular, just sitting [attached to a stone] . . . pulling in a current of water through a hole in the top and passing it out through another hole in one side [filter-feeding]" (Romer, 1968, page 137). No creature could look less like a vertebrate ancestor. But if the naturalist watched it closely, he might have seen its tiny larva, free-swimming, with the beginning of a brain in its head, a notochord and dorsal nerve cord, and a swimming tail. And—but only because the spotlight singled it out—the naturalist would have known that the improbable individual he saw carried the actual genes that, themselves evolving, went into the making of vertebrates and man.

Beyond this point I shall leave readers to use their own imaginations [guided a little by my Chapter 8, and more by such books as Romer's *Procession of Life* (1968)] as to details the naturalist might have seen under the spotlight from time to time: the first fishlike vertebrate, with internal skeleton and external bony armor, but without jaws, still filter feeding; more-familiar fishes; the lobefin coming to land; the first awkward sluggish amphibians; increasingly active terrestrial reptiles; and the first inconspicuous mammals. And the naturalist would see the environment evolve too. The continents would "drift" and would change their shapes and connections as they did so. Ice sheets would appear and disappear on the southern continents before the Age of Reptiles. And the land would become covered with forests of lycopods and horsetails, tree ferns, cycads, pinelike conifers, and finally familiar angiosperms; late in the Cretaceous the naturalist would see small mammals on the evolutionary ribbon to man living with the last ceratopsian dinosaurs in forests that included maples, oaks, and beeches in the north-temperate zone.

Vertebrate Continuities

During the 500 million years or so of vertebrate evolution, different structures, life processes, and behaviors had their own evolutionary histories within the ribbon that self-wove itself toward man. The skeleton and skull evolved; a number of bones in man's skull can be recognized in some of the most ancient bony fishes in the Devonian (Romer, 1968, page 159), and the complex evolution of the skull since then can be followed step by step. The evolution of scales from dermal bone and of teeth and hair (and, in birds, feathers) from scales can be confidently postulated if not traced in detail. The further evolution of teeth can be traced (Miles, 1972); teeth make good fossils, and they tell much about both the evolutionary relationships and the ways of life of the animals that used them. The vertebrate eye (Weale, 1978) has had its own history: early fishes "chose" to evolve a lens-and-retina eye rather than a multifacet eye like that of insects, and this choice may have been decisive in the evolution of man hundreds of millions of years later; it is doubtful whether a multifacet eye could have evolved the powers of adaptation and fine discrimination essential to human behavior and evolution of the human brain; if we could not see precisely, we might not be able to think precisely. The heart evolved step by step from a simple muscular tube to an efficient double-barreled pump. The blood evolved, complexly, as a "tissue" that carries oxygen and carbon dioxide, food and waste products, clotting agents, hormones, and antibodies (all of which have themselves evolved), and serves other functions. Thermoregulation has evolved from simply behavioral to complexly physiological. Reproduction and reproductive behavior have evolved from simple egg laying with minimal parental care to complex placental reproduction and extended care of offspring after birth. And the brain has evolved in ways described and speculated about later in this chapter.

These are only a few of the more obvious events, markers, and continuities in the complex evolutionary ribbon that unknowingly wove itself toward man from the beginning of life. The point I want to return to is that all the continuities, which can be described separately but were actually inextricably interwoven, and each of which has its own complex genetic continuity of genes that themselves evolved and interacted with other genes, ran within a ribbon which, at least since the origin of eucaryotes, has been no wider than a single species. All this (actually inconceivable) complexity has been carried by individual organisms which, if he had been there, a naturalist might have seen. This is my attempt to help naturalist-readers see man's evolutionary history as a reality since the Paleozoic, and to bring the "ribbon" of man's ancestry to a point at which later stages in the making of man can be considered in more detail.

SECTION B. TREE SHREW TO APE-MAN

The preceding pages summarize some stages and strands of evolution leading toward man from the origin of life to the beginning of the Age of Mammals. Now attention will be turned to events and processes that determined more directly that, while three or four thousand other species of existing mammals and additional thousands of extinct ones evolved with only moderate increases of brain size and intelligence, little or no use of tools, and only relatively simple social organizations, one lineage made a series of evolutionary "choices" which led to the final, rapid evolution of an animal with the unique characteristics of man.

Primates: The Arboreal Stage

Biologically, man is a primate. Primates apparently evolved from an "insectivore," which is to say from a member of the primitive placental order Insectivora, which includes shrews, moles, and several less familiar groups. The first primate may have been a "tree shrew." Tree shrews (family Tupaiidae) differ from ordinary shrews and resemble primates in having relatively large brains, eyes surrounded by bony orbits, good eyesight (as well as good sense of smell), testes in an external scrotum, and in some other anatomical and serological details. Most tree shrews are diurnal, and their activity—running and climbing—is evidently eye-oriented. Their fingers are "long and supple" and are sometimes used to hold food. And some of them are omnivorous, eating insects, other animal food, fruit, and seeds. In all these ways tree shrews seem suitable to be the ancestors of primates. But they still have claws instead of flat fingernails, and they are not social. All existing tree shrews—5 or 6 genera with about 15 species—are confined to the tropical Oriental Region, but the group may have been more widely distributed in the past. Wilson (1975, see his index) brings together pertinent information about their behavior.

   Tree shrews are unknown as fossils*, but other primitive primates appear in the fossil record early in the Tertiary, roughly 50 million years

---

* But now reported from the Miocene of India, *Nature* 20 September, 1979, 213–215.

ago. The fossils have been found only in what is now the north-temperate zone, but primitive primates were probably also in the Old-World tropics then. Some of them were or became lemurlike. And one or more "lemuroids" evolved to become monkeys. The actual course of their evolution was probably complex but is unknown in detail. We do not know, for example, how tarsiers (now surviving only on tropical islands southeast of Asia) fit into the evolutionary history of primates, and we do not know either the ancestor or the geographic history of South American monkeys, whether they evolved separately from a primitive primate or are derived from the same ancestor as the Old-World monkeys, and if the latter, whether they reached South America by way of North America or directly from Africa. But these details are beside the point. In any case, monkeys were in the Old-World tropics by the Lower Oligocene, perhaps 40 million years ago.

The evolution of arboreal primates included a number of adaptations that turned out to be pre-adaptations in the evolutionary fabric weaving itself toward man. The ancestral insectivore was probably a ground runner and claw climber, as most tree shrews are now, but its lemurlike and monkeylike descendants evolved a new set of arboreal adaptations. These included grasping hands and feet, with flat nails instead of claws and with opposable thumbs; the immediate selective advantage of hand climbing instead of claw climbing probably was that it allowed an increase of body size. The sensitivity of the hands, and the area of the brain that controlled their sensitivity and dexterity, increased; the hands then replaced the tactile bristles that most mammals use to feel their way in close quarters. Hands like these were adapted to holding on not only to branches but also to each other (especially mother and young) and to food, and were pre-adapted to the precise manipulation of human hands. The change from claws to nails *may* have been because, in jumping from handhold to handhold, clawed hands did not let go of branches as cleanly as flat-nailed hands did, so that clawed individuals were checked in mid-jump, fell, and were selectively eliminated; and flat nails may have proved more useful than claws in manipulating small objects. Roughening of the grasping surface of the hands, by formation of minute ridges, dates from this time. It increased the security of holding—and, now, makes the fingerprints we use for individual identification.

The arboreal stage saw also profound changes in the roles played by nose and eyes, and in the eyes themselves. Most insectivores, and most mammals, have much greater powers of scent than we do: their behavior is more odor oriented, although they can see too. But odor is not so precise a guide for leaping from one branch to another as sight is, and

our arboreal ancestors, from tree shrews (if they were ancestral) on, became increasingly dependent on sight and less on scent. This was probably accompanied by a shift from nocturnal to diurnal activity, if the
pre-primate ancestor was nocturnal. Some other arboreal mammals including some primitive primates are nocturnal, but the advantage of
eye-oriented behavior is greater by day than by night, and most tree
shrews, monkeys, and apes have "chosen" this advantage. The evolution
of binocular vision—the change of position of the eyes so that they both
focus on the same objects—occurred during this time; it gives an additional advantage in judging distances. And color vision too evolved or
was improved; it increases power of discrimination of details for eye-
oriented animals. Campbell (1974, page 333) suggests that these changes
made perception more three-dimensional. I am not sure about this—
even a ground-living shrew must perceive three-dimensional space—but
arboreal life probably did result in more precise visual perception of the
environment.

Some of these details are speculative, but man's evolution later did
depend on eyes that could judge distances and discriminate details at
arms length, and on hands capable of precise dexterity. We surely do
owe much of our superb physical equipment for human skills to ancestors that clung, and swung, and jumped among the branches in tropical
forests in Africa or southern Asia tens of millions of years ago.

Many human behaviors may trace from man's arboreal ancestors.
Most obvious may be vocal communication. Calling back and forth presumably helped to keep families and, later, larger social groups together
in dense forest canopies; increasing the variety of calls would give increasing amounts of useful information; and calling might be less likely
to attract predators in the trees than on the ground. Monkeys do vocal-
ize more than most animals—witness the noise in monkey houses in
zoos—and this may well have begun the evolution of still more complex
signal and communication systems and finally language.

Other arboreal adaptations that turned out to be pre-adaptations
probably included a close mother-child relationship; females have to
hold their young especially securely when moving through the tree
tops. Vestiges of this behavior may still be visible in human mother-
child bonds, and, more important, the close relationship of females and
their young during the arboreal stage pre-adapted man's ancestors to
evolve still closer family bonds that were, later, the basis of human social organizations. And single births—bearing one young at a time—was
perhaps an arboreal adaptation which focused the attention of parents
on one offspring at a time and favored effective teaching.

Social organizations evolved complexly among arboreal primates. They have obvious direct advantages in exchanging information and in cooperative activities—finding food, detecting predators, and fighting competitors—and socialization had effects that are less obvious but probably very important in the evolution of man. One was an increase in altruistic behavior. Nonsocial animals are selfish except, temporarily, in their relations with their mates and offspring. Social organization greatly increases what individuals do for each other; if what they do is indirectly selfish (if altruism is a net-gain lottery; see Chapter 7), it is still altruism, emotionally and behaviorally. Altruism is itself essential to humanness, and some altruistic behaviors allow delay in response that may favor evolution of a delayed-action, thinking brain (this chapter, under "Evolution of brain and intelligence"). And social organization presumably was accompanied by evolution of dominance systems (often too narrowly called "peck orders"); having a dominant leader presumably helped a group of social primates to act effectively; and the emotions that go with dominance systems may still affect our individual behavior, the forms of government we accept, and perhaps even our philosophies (Chapter 10).

Also—another matter—it has even been suggested that man's carelessness with litter may come down from arboreal ancestors, who could drop food remains and feces with less risk of betraying themselves to predators and less risk of transmitting diseases than ground-living animals can.

This is an absurd—but necessary—oversimplification of what is known now about primate behavior and its evolution. Wilson brings much of it together in *Sociobiology*, especially in Chapter 26 and Figure 26–1. He notes that tree shrews, tarsiers, lemuroids, New-World monkeys, Old-World monkeys, great apes, and man form a sequence of grades of increasing anatomical specialization, behavioral complexity, and social organization. But, as he warns, this is not a simple evolutionary continuity; primate evolution has surely involved continual geographic and ecologic radiations and reradiations, parallelisms and convergences, and extinctions as well as adaptive advances; the sequence of grades is only the net result of these complex and largely unknown events. Nevertheless a one-species-wide evolutionary ribbon, represented by individual animals that a naturalist might have seen, ran from tree shrews (if they were ancestral) to man, weaving into itself all the step-by-step primate adaptations of behavior as well as of structure that turned out to be pre-adaptations to the final making of man, and it is legitimate, fascinating, and important to try to trace the ribbon's history.

Apes: Brachiating

Although we have only a few actual details, we do know that during the Tertiary Old-World monkeys diversified in tropical Asia and Africa, where many of them still survive, and that some of them became ape-like. Pertinent fossils are few and their significance is doubtful, but *Dryopithecus* in the mid-Miocene and *Ramapithecus* in the late Miocene and early Pliocene may have been ancestral hominids. Both are known fossil in southern Asia and Africa. They had incipiently hominid dentitions (U-shaped arcades, with canines less prominent than in most monkeys); they may have been brachiating arboreal animals which were partly ground living and capable of some two-legged walking; and *Ramapithecus* may possibly have used stones as tools. For further details, many of which are argued about, see Leakey and Lewin (1977), *Origins*. I have not tried to fit into this history still-more-recent discoveries reported in the press. Their significance may be exaggerated.

The "apelike" animals were tailless, larger than most monkeys, with relatively longer arms which adapted the animals to "brachiating," or to swinging or throwing themselves from handhold to handhold rather than jumping from branch to branch. Brachiating is sometimes called "arm swinging," but this phrase has another more common meaning. It has been called also "suspensory progression," and might (I think) better be called hand-traversing. Apes still get around in trees this way. The evolution of brachiating was accompanied by strengthening and stiffening of the back, broadening of the pelvis, and a tendency to place more weight on the hind legs and feet than on the front ones when walking. These changes are not all shown by the inadequate fossils, but are in part inferred from the structure and behavior of existing apes. They presumably brought the brachiators an immediate advantage, which may have been primarily a further increase of size. A heavy body can be moved more safely through the treetops by using successive handholds than by jumping, and safety from falling becomes increasingly important with increase in size, not only because of the increased danger of branches breaking, but also because of the increased likelihood of falls resulting in serious injury.

Brachiating and the changes that accompanied it pre-adapted an ape in several ways to return to the ground and to take the successive behavioral and anatomical steps that led toward man. A fairly large size probably was itself an essential pre-adaptation; a much smaller animal could hardly have evolved as a stone-throwing hunter pursuing large game in the open, and perhaps could not have evolved a human brain. The changes in back and pelvis and the shift of weight toward the hind legs and feet pre-adapted the animal to begin to stand erect and to walk on two legs. And the long, loose-jointed arms pre-adapted it to begin to hit and throw. Human beings, especially children, still brachiate in a modest way, on overhead ladders if not from branch to branch, and still enjoy it.

The increase of size that brachiating permitted necessitated more time for young to reach maturity, which later meant more time for teaching them.

Apes: The Descent to the Ground; Knuckle Walking -

In the course of time—after perhaps 40 million years of evolution in the trees—an ape on the evolutionary line toward man returned to the ground, bringing with it adaptations primarily for arboreal life and secondarily for brachiating which pre-adapted it for further evolution of behavior and structure. The move to the ground was probably not made suddenly but gradually, and the initial advantage of it probably was that it allowed food resources on the ground to be added to those in trees. Chimpanzees still get food from trees and from the ground, and so in fact does man.

This was not the only primate to return to the ground. Others that did so included the ancestor of modern baboons, and baboons, like man's ancestor, spread from the forest to more-open country. But the baboons remained baboons, while a secondarily ground-living ape evolved to become man. This was, I think, because the ape brought with it pre-adaptations inherited from brachiating ancestors while the baboons did not, but a secondary detail of behavior may have been more directly decisive, as suggested later in this chapter.

Just what the ancestral animal was that returned to the ground is not known. There are two modern ideas about it. One is that it was not an ape, or at least not directly related to the existing great apes, but a separate stock. Proponents of this idea sometimes say that man is not descended from an ape, but this is obfuscating semantics; whatever the animal was, it surely looked like an ape and would have been considered one by any nontechnical observer. The other idea, consistent with protein similarities and with nonadaptive anatomical details (i.e., with details not changed recently by divergent ways of living and doing), is that man and the existing Chimpanzee and Gorilla are closely related and derived from a common ancestor who lived perhaps as much as 10 or as little as 5 million years ago.

The common ancestor was probably a knuckle walker; it walked with its weight chiefly on its hind limbs, with the hind feet still handlike but placed almost flat on the ground, with the toes bent under only a little if at all; the forelimbs bore some weight too, but with the fingers folded under and the knuckles on the ground. This posture was dictated by the extra length of the arms, inherited from the brachiating ancestor. It is an awkward way of walking. Chimpanzees and gorillas still do it when walking slowly, but are not agile walkers. But it was good enough to allow an arboreal ape to move over the ground, and it proved to be a double pre-adaptation. Putting the animal's weight mainly on its hind limbs favored evolution of a more erect posture and two-legged walk-ing and running; even chimps can run two-legged, awkwardly. And put-ting the knuckles rather than the flat of the hands or the finger tips on the ground allowed the animal to carry objects in its hands while mov-ing on the ground. This may have been a small but decisive detail in the evolution of a carrying, hitting, throwing, tool-using animal. It may have been a principal reason why a ground-living ape became man while ba-boons, which walk on the fingers of their hands, did not. The knuckle-walking posture is still, incidentally, fairly comfortable for man; football linesmen are said to use it.

Just when the move to the ground occurred is not known. It might have been anywhere from 5 to 10 million years ago, more or less. We would like to know about this, and we would like to know just what the animal was that made the move. But lack of knowledge about these details should not be allowed to obscure the more fundamental things we do know or can infer about the successive adaptations and pre-ad-aptations made by apelike primates that prepared one of them for the final evolution of man.

Apes: The Spread to Open Country

Apes, some of them ground-living knuckle walkers, diversified to some extent in the Old-World tropics (but fossils are few) and, probably in the Pliocene perhaps less than 10 million years ago, one of them began a series of giant evolutionary steps that led to man. The steps were eco-logical, behavioral, and anatomical, all inter-related.

The giant step ecologically, which probably initiated the others, was a movement—or more likely an extension—from forest into more-open country. It has been suggested that the move was forced by reduction of forest caused by decreasing rainfall, but this seems unlikely or at least an oversimplification. Generally, successful, evolving animals move to obtain advantages more than to escape disadvantages. The move probably began gradually, as an extension of the animal's ecologic range to include more-open areas—chimpanzees have extended their ranges a little in this way—and the extension probably brought an immediate net advantage. It added the food resources of open country to those of forest, without adding too much to the risk of living. If new predators were encountered in the open, forest-living predators and forest diseases were left behind. And the extension was perhaps at first only into open places with scattered trees in which apes could take refuge from ground-living predators and at night. Baboons often do this, and so, on occasion, do men.

The giant step behaviorally was to begin to stand more erect and to walk and run on two legs. This too presumably brought an immediate net advantage. It increased the speed of running while leaving the hands free, and it increased the range of vision. An ape in the forest has less need to see far, and cannot see far if the foliage is thick, but in the open it is possible and advantageous to extend the distance at which both food and enemies can be seen and recognized. The hand and arm were probably used increasingly for carrying, hitting, and throwing during this time. Carrying increased in importance with the increase in distances covered, and throwing became more effective in open country. Some short-range throwing, the kind that chimpanzees do (awkwardly and not very effectively), may be useful in the forest, but to throw accurately at longer range is possible only in the open. Try throwing stones at a target 30 meters away in thick woods!

Although the two-legged ape increased its running speed, its structure did not allow it to equal the speed of the larger four-footed mammals that were prey or predators. Perhaps in compensation for this inherent disadvantage, the ape evolved the power of running at moderate speeds but for longer times. In effect, it made a forced evolutionary choice: persistence rather than speed. This *may* have been accompanied by an increase in attention span. A notable difference between young apes and young humans is that the latter habitually focus their attention on single activities much longer than apes do, and this is important to human learning. This is an inherited genetic difference, which must have had a beginning and an evolutionary history, and it *may* have begun as a selectively advantageous habit of keeping the end in mind during longer and longer periods of running. Of course this is only a guess, and surely an oversimplification.

The principal disadvantage of standing and running presumably was that the ape body, although pre-adapted in a general way, was imperfectly adapted in detail, and the imperfections must have resulted in frequent incapacity and death. A secondary behavioral disadvantage of the more erect posture was that it necessitated a change from back to front copulation. This was accompanied or followed by complex changes in sex signals which may have confused our evolving ancestors as much as sex confuses some humans now. Campbell (1974, especially pages 298–299) has more to say about this.

The giant step anatomically was to begin to adapt the ape body to the new posture and new mode of locomotion. This involved changes in the skeleton: a stiffer back and broader pelvis, changes in the shape and articulation of the leg bones, changes in the angle of articulation of the skull to the spine, and the beginning of evolution of a walking and running foot. These skeletal changes are indicated in the fossil record, which shows them well advanced, although not complete, several million years ago, long before the dramatic evolution of the human brain. Apparently our ancestors stood erect or nearly so and walked and ran on two legs for several million years before they began to think as we do. The more erect posture and longer-distance running must have required changes also in the heart, which had both to move the blood a greater distance vertically and meet the added demands of sustained activity. The hand and arm evolved a better carrying, hitting, and throwing mechanism. The eye too may have changed, toward increased power of adaptation to distant as well as close vision. And the thermoregulating system probably changed complexly as discussed later in this chapter.

Of course these changes all occurred by variation and selection. Selection occurs by elimination (Chapter 6A), and the evolutionary changes listed in the preceding paragraph must have involved the premature crippling, death, or failure to reproduce of untold millions of individuals with weak backs, weak hearts, poor eyes, or other inferior characteristics, including inferior behaviors and inferior brains. These selective processes are still continuing in man, although reduced in force. The evolutionary or genetic load carried by the ape that spread into open country and began to stand erect was heavy. But the cost of not changing would presumably have been still heavier. The advantages outweighed the disadvantages.

The Naked Ape

Man differs from most mammals and from all other primates in his nakedness. This fact suggests that our nakedness is recent, and other evidence supports this suggestion: the human foetus is well furred, but usually sheds its furry coat before birth; and the skin of adult humans is still covered with hairs—as many as cover the skin of well-furred apes of the same size—but the human hairs are so short and fine as to be hardly noticeable, except where specially developed. This is strong evidence that our ancestors had a coat of hair and that they "chose" to decrease the quality of the coat but have not disposed of it during evolution.

The primary advantage of nakedness to our ancestors probably had to do with temperature regulation. Animals running in the open on hot days *must* get rid of surplus heat. Other running mammals in Africa have not gone naked, but some of them run mainly at night, or run for only short distances, and all are descended from long lines of running ancestors and have had time to evolve profound adaptations for temperature regulation. For example, some running carnivores have a heat-exchanging, brain-cooling network of blood vessels at the base of the brain that man lacks (Baker, 1979). Man's ancestors were slow-moving forest dwellers who ran very little, and when they came out in the open and began to run for increasing distances, they had to evolve heat-losing mechanisms quickly from whatever materials were available. For them the quickest available way was to increase heat loss from the skin by reducing the coat of hair and at the same time increasing sweating. Man does sweat much more copiously than other primates. Perhaps our ancestors made an evolutionary choice at this time: to cool themselves by reducing hair and sweating rather than by hanging out their tongues and panting. The hair-reduction-and-sweating mechanism may have been quicker and easier for an ape to evolve. And perhaps the tongue was already committed to another use; to vocal communication.

Man usually thinks of his nakedness as a peculiarly human characteristic, but this idea is probably more emotional than scientific. The need of evolving quickly an improved cooling mechanism probably coincided with the spread into open country, and our pre-human ancestors may have been as near-naked as we are from that time or soon afterward.

Cooling is only one of the probable selective advantages of nakedness. The net advantage is complex. Against it is that it exposes the naked animal to inclement weather, and makes entry into colder climates impossible without fire and clothing. It also exposes the skin to ultraviolet radiation, of which some is beneficial but too much is damaging; this difficulty was apparently countered by development of skin pigment in the tropics, and later reduction of it in the less-sunny north-temperate zone. Loss of protective hair probably also made the naked ape more susceptible to injury in attack by predators and in fighting. On the other hand, reduction of hair reduced the cover under which annoying and disease-carrying ectoparasites could hide, and it made the skin more available for social signals. We use, and our ancestors probably long have used, facial expressions to communicate states of mind and intentions; Darwin had a lot to say about this. This matter is complicated by sexual dimorphism: the almost hairless female face may be adapted to communicating with young as well as with adults, while the facial hair of the male may be more adapted to advertising maleness and to threat signals.

SECTION C. APE MAN TO MAN

This evolutionary history has now reached the point at which an ape had spread from the forest into more-open country and had begun to evolve a posture, anatomy, physiology, and behavior (all interrelated) that pre-adapted it to become man. But the animal was still pre-human, one species of mammal among hundreds of others that existed in Africa, if this is where it was. Now to be considered is the evolution of the pre-human ape into man.

Up to this point man's evolutionary ribbon can be traced only by grades or adaptive stages, at each of which radiations occurred. We cannot identify man's actual ancestors within the radiations, and even the most informative fossils are more likely to represent collateral relatives than actual ancestors. But from this point on it is first reasonably possible and then probable that known fossils are the remains of ancestral individuals whose genes, with very little further evolution of their own, wove themselves into man.

The Fossil Record: Man Apes to Man

From the mid-Tertiary into the early Pliocene the fossil record of pre-human primates (briefly summarized under "Apes: Brachiating" in Chapter 9B) is scanty but does suggest the appearance of hominids—apelike primates with incipiently human dentition—in the Miocene. In the mid-Pliocene there is a gap in the record, which we would especially like to fill! But in the late Pliocene and Pleistocene the record is extensive (and being rapidly added to), and many of the fossils have now been dated by methods which have only recently come into use and which have led geologists to place the beginning of the Pleistocene about 3 (rather than 1) million years ago. Many of the fossils are fragments, but they include also a number of skulls and other significant bones, and they have been thoroughly studied. Pilbeam and Gould (1974) say that "Human paleontology . . . contains more practitioners than objects for study. This abundance of specialists has assured the careful scrutiny of every bump on every bone." It has also assured an abundance of divergent conclusions and disputes. I shall not try to catalogue either the bones or the disputes, but shall simply summarize what seems to me to be the essential history that fossils reveal. In doing this, I shall first follow Pilbeam and Gould, whose paper reflects both a thorough knowledge of the fossils and common sense, and shall then indicate how recent finds have changed the history.

The pre-humans in question are known as Plio-Pleistocene hominids, more informally as man apes or ape men, and formally as species of *Australopithecus* (Australopithecinae or australopithecines) and early *Homo*. They were not as big as we are, but were comparable to existing pygmies in size. They apparently stood nearly erect, and walked and ran two-legged. And they apparently lived in open country. But they still (at first) had brains not much larger than those of apes.

The earliest australopithecine fossils are in Africa and are between 5 and 6 million years old. They represent a man ape who was relatively lightly built and still rather small brained, and who is usually called *Australopithecus africanus*. From this ancestor or something like it two lineages evolved divergently. In one, the *robustus* or *boisei* lineage, although the body size increased, the brain and teeth stayed in scale (see Pilbeam and Gould for a discussion of "scaling," which is not a simple matter). The animals in this lineage were probably vegetarian-omnivores; they had teeth that suggest a diet "including small tough objects such as seeds, roots, tubers, small animals, and so forth." This lineage apparently became "extinct without issue" before the end of the Pleistocene, about 1 million years ago. In the other, the *africanus* lineage strictly speaking, the size increased too, but the brain increased in size even more rapidly—much more than scaling required—and the dentition weakened, suggesting a softer diet. This lineage evolved to become man. A disputed detail is whether *Australopithecus* evolved entirely in Africa or occurred also in southern Asia, where only doubtful fragments have been found; my guess is that the theater of evolution was the entire accessible area of the Old-World tropics.

Recent fossil finds (well described by Richard Leakey and Lewin, 1977) force three modifications of the preceding history: *Homo* is older than previously supposed and coexisted with *A. africanus* for a considerable time; several additional species of *Australopithecus* and perhaps even of *Ramapithecus* coexisted with early *Homo;* and the primary theater of diversification of *Australopithecus* and *Homo* was apparently Africa, the spread of *Homo* to southern Asia occurring later. But I think the new evidence does not invalidate Pilbeam and Gould's admittedly simplified description of divergence of *Homo* from an *africanus*-like ancestor, but only puts the divergence earlier, and adds branches to the main phylogenetic tree.

Some recent finds, including "Homo afarensis," are reviewed by Gould (1979). "Molecular clocks" indicate that the pre-human lineage separated from the common ancestor of gorillas and chimps perhaps only 5 or 6 million years ago, and Gould suggests that the evolution of posture from knuckle walking to upright running may have been as rapid as, 5 or 6 million years later, the final explosive evolution of the human brain. I think he is probably right about this.

The pre-human lineage—Homo in a broad sense—apparently began to diverge from other hominids at least 5 million years ago and had become Homo habilis 2 or 3 million years ago. It was a subhuman hominid with a relatively larger brain and smaller teeth than coexisting Australopithecus, although it was still far short of the structure of modern man. Homo habilis is African, but the African population need not have been directly ancestral to man. More likely it represents a grade or level of pre-human evolution. Populations of more or less this grade were probably widely distributed in southern Asia as well as Africa within a time span of about 2 million years, from (very roughly) 3 to 1 million years ago. The populations probably diversified in both structure and behavior within this time, but we have no clear record of the details.

The next recognized pre-human evolutionary grade is represented by Homo erectus, the famous fossil ape man of Java, who was still small-brained compared to modern man. Populations of more or less this grade were widely distributed in southern Asia and Africa, from (again very roughly) 1.3 to 0.3 million years ago.

Finally, fossils show a sudden, rapid increase of brain size in man's ancestry beginning apparently only about half a million years ago—in about the last one-twelfth of Australopithecus/Homo history—leading rapidly to Homo sapiens, and then a transition from a big-jawed to a small-jawed form of sapiens about fifty million years ago.

Sites, Refuse, Tools, and Fire

Places where ape men and early man lived and worked, refuse deposits, primitive tools, and evidences of fire tell a good deal about our possible or probable ancestors.

The best preserved sites are caves, which sometimes contain bones of prehumans as well as stratified deposits of their refuse, including plant fragments, mollusk shells, and animal bones.

Tools may have been used by pre-human apes long before they become recognizable in the actual record. Chimpanzees now use sticks as clubs and levers, throw sticks and stones in play and as weapons, and use handy objects as tools in other ways. For example, a chimp will sometimes break a hole into a termite mound, prepare a twig by stripping off the leaves, push it into the mound, pull it out, and eat the termites clinging to it. Wilson (*Sociobiology*, pages 172–175) lists nine categories of tool use by chimpanzees, and Jane Goodall (van Lawick-Goodall) describes her own observations in *In the Shadow of Man* (1971), a fine book written by a naturalist and exciting reading for other naturalists. Man's ape ancestors may have used such simple tools for several million years, transmitting their use by imitation.

The first actual evidences of tool use are accumulations of "pebble tools"—small stones worn by use or broken to sharp edges. These were apparently used by early *Homo* at least $2\frac{1}{2}$ million years ago, and perhaps much earlier. The first use of broken stones may have been to cut up large animals with skin too tough for an ape man's teeth to bite through.

Tools made first of crudely broken and then of more and more precisely chipped and flaked stone show the later evolution of stone technologies. Their evolution was much more complex than can be described here, but very slow; for at least $2\frac{1}{2}$ million years man's ancestors gradually improved their chipped- and flaked-stone tools, and used bone and shell (and probably wood, which is not preserved) in simple ways, but made no dramatic innovations. The first great breakthrough occurred when man began to give stone tools smooth surfaces and precise working edges by grinding them with other stones instead of chipping them. This marked the end of the Old (Paleolithic) and the beginning of the New (Neolithic) Stone Age. The invention of the bow may date from about this time. Stone grinding began perhaps less than 10,000 years ago, and was immediately followed by the beginning of agriculture and the rapid evolution of increasingly complex technologies, communities, and modern civilizations. The period of accelerating social evolution, from the beginning of stone grinding to the present, was only about $\frac{1}{250}$ as long as the time span within which pre-man and man had so slowly improved broken, chipped, and flaked stone tools!

Man-made hearths and other evidences show that *Homo* had fire more than half a million years ago in Europe, and less formal uses of fire may have begun earlier. Krutch thinks that "domestication" of fire was the bravest thing man ever did. And *domestication* was probably what it was at first—fire taken perhaps from lightning strikes, and tamed and fed. Fire *making* may have come much later; we have no way of knowing when. The extinct Tasmanians, who had probably the most primitive of recently existing cultures, apparently kept fire but did not know how to make it. But perhaps they had forgotten something that their ancestors knew.

Fire was useful in several obvious ways: for warmth, to protect camps against big carnivores, to drive game, to harden wooden spear points, and for cooking. Coon, in *The Story of Man* (1954; still a book that naturalists can read with pleasure and profit, if they remember that many details have been added to the record since then), estimates that when man ate roots, seeds, and meat raw, getting and eating food took most of his time, but that when he began to cook food, eating time was reduced to about 2 hours a day. Whether or not this estimate is accurate, cooking did allow division of labor—1 person could cook for several—and provided free time that could be used to make tools, teach children, experiment, talk, and think. Leakey and Lewin (in *Origins*, pages 151–152) suggest an even more important use of fire. Camp fires drew people together, facilitated exchange of information, and focused social activities. The pleasure we feel in companionship before a fire comes down to us from very early times!

The Extension Northward

For several million years evolving *Homo* was held within the warmer part of the Old World, but toward the end of the Pleistocene populations began to extend northward. The earliest actual records are of *Homo erectus* in Europe and China 500,000 and 300,000 or more years ago respectively. Hearths and burned bones at living sites show that *Homo erectus* had fire and probably hunted and surely cooked and ate large animals, although other food must have been gathered too.

It is tempting to speculate that the northern climate required invention of new facilities and technologies, especially fire-related ones, and that modern, primarily fire-dependent technologies and cultures evolved from these beginnings entirely in the north. This may have been the case, but it does not follow that the center of evolution of man shifted northward. Different races can and do appropriate each others' cultures. The probabilities seem to me to be that the evolution of *Homo erectus* into *sapiens* continued in southern Asia and Africa; that successive northward extensions occurred, with perhaps occasional smaller counter-movements southward; and that each new northward-extending population replaced or mixed with the preceding one, and appropriated its technologies. Neanderthal Man may have represented one such population, Cro-Magnon Man, another, but these details only hint at the complexity of dispersals, mixings, and replacements that may have occurred in 500,000 years. This is, I think (thinking as a zoogeographer), how dominant animals do evolve and disperse.

The present situation is consistent with this speculative history. Northern races now have an annihilating technology, but they seem more likely to use it against each other than against more southern populations; existing Old-World tropical populations are, biologically, fully evolved *sapiens*; and movement *of people* now is apparently more from the Old-World tropics northward than the reverse.

Human Races and Their Evolution

Races of man are comparable to geographic subspecies of other animals, described in Chapter 6C, and the ethical problems that races pose for modern man are considered in Chapter 10. In the present chapter I want only to say what naturalists who try to be objective can see.

Races of man can be seen to exist. The main ones occupy broad geographic areas; minor ones are ethnic groups which have more limited geographic origins. It is true that all human races share most of their genes and are interfertile, that they are genetically complex within themselves, that they are imprecisely bounded and often grade into each other, and that their compositions and boundaries are continually changing, but these are expected characteristics of subspecies of other animals.

The major races of man listed and mapped by Coon (1962) are Australoid, in Australia and isolated areas in the East Indies and India; Mongoloid, northern and eastern Asia and the western East Indies, the Americas, and the islands of the Pacific; Caucasoid, Europe and northern Africa east to India; Congoid, the main part of Africa; and Capoid, South Africa. Subdivisions of the races can be recognized, but are argued about and will not be considered here.

Although human races clearly have evolved, and have differentiated partly by adaptation to different environments, their actual histories are obscure. Coon (1962, page 657) proposes as a thesis that

"At the beginning of our record, over half a million years ago, man was a single species, *Homo erectus,* perhaps already divided into five geographic races or subspecies. *Homo erectus* then evolved into *sapiens* not once but five times, as each subspecies, living in its own territory, passed a critical threshhold from a more brutal to a more *sapient* state."

This thesis has been attacked, and Coon accused of racism because of it, with a violence that ought to have no place in science.

Although existing human races probably originated more recently, I can suggest a modification of Coon's thesis that makes it biogeographically acceptable. For several hundred thousand years mankind may have evolved within the main part of the Old World as one species within which populations differentiated in partial isolation. But the isolation was only partial; continual dispersals, counterdispersals, and mixings may have occurred which tied the populations together, not enough to prevent their partly independent evolution, but enough to limit their divergence. I see nothing racist in this modified thesis. It allows us our common humanness and also our human individualities.

The Hunting Hypothesis

Chimpanzees, but not gorillas, occasionally kill and eat small baboons and antelopes, and form briefly organized hunting groups to do it. When a suitable victim happens to come close, several chimps surround it, take positions apparently calculated to cut off its escape, kill it, and share the meat. Our ape man ancestors may have begun like this several million years ago. And cave paintings a few thousand years old, and existing hunter-gatherers, show the other end of what was presumably a long process of evolution of hunting-group behavior. But we have little actual evidence of what happened between the postulated beginning and the observed end of this inferred evolutionary continuity. Animal bones found at Plio-Pleistocene living sites are usually not significant, for there is usually no evidence that the animals were killed and eaten by hominids. Exceptions are a site where a number of baboon skulls were found fractured in a consistent way, as if the animals were killed with clubs and brought in to be eaten and, at later sites, bones marked by fire and broken as if to extract the marrow.

However, the hunting hypothesis is explanatory:

Hunting, with the eating of more meat and less hard food, and using weapons instead of teeth in fighting, explains the weakening of the teeth in *Homo*.

Hunting explains the move from forest into more open country, where animals to be hunted abounded.

It helps explain the selective advantage and evolution of throwing, as a hunting technique.

It explains the beginning of the use of broken stones as tools, if first used to cut up large animals.

It helps explain the selective advantage and evolution of organized groups with effective, dominant leaders.

By favoring throwing, tool use, and organization, hunting helps explain the evolution of a complex, thinking brain.

And it gives a reasonable explanation of the success of *Homo* in competition with other hominids, as a result of better use of resources, more effective fighting, and increased intelligence.

So the hunting hypothesis helps explain the divergence of *Homo* (the aggressive, hunting, meat-eating omnivore) from *Australopithecus* (the less aggressive, vegetarian omnivore) and later the competitive elimination of the latter.

Two more points: the pleasure men take in hunting suggests reinforcement of a behavior that has been strongly, selectively advantageous. And it has even been suggested that men and dogs understand each other so well because both are derived from ancestors that hunted in groups.

The hunting hypothesis is not new. Read, in 1925, argued persuasively that "man was differentiated from the anthropoids by becoming a hunter" and refers to pertinent publications back to 1900 and to hints by both Darwin and Wallace. This hypothesis is mentioned again in Chapter 10.

Judging from chimpanzees now and from what we know and infer of our own history, hunting by our ancestors was a group activity. Hunting behavior therefore evolved partly by group selection, which (as always) can be translated into terms of individual selection: individuals who acted effectively both as individuals and as members of hunting groups survived; their groups survived; and less cooperative groups were selectively eliminated, probably often by group fighting and small-scale genocide. This probably began far back in pre-human history, was widespread within historic times, and still continues. It is the root behavior from which modern warfare has evolved. Modern humanitarian man does not like to think this of his ancestors: that war has roots in pre-human and early human societies and has been an effective mechanism accelerating their evolution. But surely if we hope to control war it is better to look squarely at the facts and probabilities than to try to hide them behind a humanitarian screen. Hiding destructive biological and evolutionary forces does not inactivate them! But if we can recognize them as outmoded heritages from the past—part of the heavy evolutionary load we carry—we can perhaps control them. Ordinary people may have better judgment about this than some activist biologists.

Sex, Family, and Social Groups; Altruism; Dominance Systems

The evolution of human sexual behavior, family groups, and early social groups can be reconstructed only by inference, but the inferences are sound and very important to understanding the final steps in the evolution of man.

All mammals mate, but most of them do so only at special times, and the females then bear young, give them milk, and protect them at least briefly. But some mammals do more than this; female parents and young, and sometimes male parents too, stay together for a while in family groups, and families sometimes join in larger social groups.

Among primates, although sexual behavior is diverse, sexual activity is usually not continuous but occasional or casual; females care for their offspring for extended periods, but males usually do not; and social groups are diverse, as described by Wilson in *Sociobiology*.

The first step toward the special human social organization was apparently that females became continuously rather than occasionally attractive to males. This tied males into stable family groups. And this resulted in better care of offspring, increased division of labor, and increased opportunity for learning and transmission of learning by imitation and, later, by deliberate instruction. These advantages were increased by an increase of the time during which offspring were dependent on their parents, and perhaps by lengthening of the attention span during which individuals kept their brains focused on particular ideas. And the advantages were presumably further increased when relatively stable families joined in larger social groups.

This is, of course, a simple outline of what must have been a very complex history, which was (I think) at first genetically determined but later became increasingly a process of cultural heredity and cultural and social evolution, with the genetic mechanism overlaid and perhaps reduced (Chapter 10). The essential reproductive behavior of males and females was surely genetically established far back in vertebrate history; it has been continually modified since then, but never lost. The increased sexuality of man's immediate ancestors surely was genetically determined too, and so were the increased time of dependence of offspring and the lengthening of the attention span. But beyond this it is difficult to separate genetic and social determinants in man's sexual, family, and social behavior. It is probably safest, and nearest the truth, to think of all man's nonessential social behavior (i. e., all social behavior not directly essential to reproduction) as having a general genetic base which allows diverse modification socially. This is probably true of division of labor between the sexes; there evidently are genetic differences beyond the obvious essential ones, but they can be and are over-ridden socially; they are then comparable to outmoded vestigial organs. I see nothing in our evolutionary history inconsistent with equality of rights and opportunities between the sexes. But I suspect that many individuals will still choose according to their sex chromosomes rather than their social rights and opportunities!

In any case, increasingly complexly organized social groups evidently did evolve in the pre-human lineage, and their evolution not only had the advantages just suggested, but also greatly increased the roles of altruism and dominance systems in pre-human and human behavior.

Altruism is inherent in social organization. It is an essential characteristic of man, with a general genetic base inherited from our social ancestors, although the details of our altruistic behavior now do not seem to be genetically determined (Chapter 7).

Dominance systems, in which some individuals take precedence over others in obtaining territories, food, or mates, are conspicuous in many animals including many primates. They are genetically determined: dominant individuals are not genetically differentiated, but each individual in the system inherits *both* a compulsion to dominate when it can *and* a compulsion to submit when it cannot. (Changes of dominance therefore can and do occur; in man, a submissive corporal can become a dominant dictator.) Dominance systems are evidently selectively advantageous to many nonsocial species—they allow some individuals to survive and reproduce when resources are scanty or competition intense—but their advantage increases in social groups, which gain by acting as units under dominant leaders. Dominance systems were presumably increasingly advantageous as man's ancestors became increasingly social, and are conspicuous and perhaps essential to many human activities. This is simple enough in theory, although exceedingly complex in fact. But note; an effective dominance system does require that individuals have not only a compulsion to dominate but also a compulsion to submit, and the latter may not only facilitate dictatorships but may also be the basis of human philosophies and religions (Chapter 10).

SECTION D. BRAIN, INTELLIGENCE, AND CONSEQUENCES

Evolution of Brain and Intelligence

A large, intelligent brain is *the* essential characteristic of man. I shall look at it and its evolution is several ways, some of them highly speculative.

The brain has two overlapping functions: it (1) monitors and controls the animal's internal, largely unconscious metabolism and homeostatic mechanisms, and (2) makes models of the animal itself and of the things around it, relates them to each other, makes predictions, and determines behavior. Thinking and intelligence are products of this second function of the brain; intelligence can be defined as proportional to the power of recognizing and integrating complex factors, comparing alternatives, and making effective predictions and decisions.

Through most of vertebrate history selection presumably favored a brain that focused on major, immediately important objects and events (i. e., other animals, things big enough to eat or to be dangerous, and events that directly affected the animal's needs or safety) and that responded instantaneously, producing a limited "repertory" of useful stereotyped behaviors. (Readers should remember that selection could "favor" such a brain only indirectly; its evolution had to wait for successive random genetic events that made better brains of the kind described, each genetic improvement being followed by costly eliminations of great numbers of individuals with other kinds of brains, including brains that perceived too many details too precisely, or did not respond promptly.) From such a coarse-grained, instantaneous brain a more fine-grained, delayed-action brain capable of mammalian intelligence and human thinking evolved in situations in which the direction of selection was partly reversed, favoring finer discrimination, more complex integration, and delay in some responses, allowing time for comparisons and choices among alternatives. (Evolution of a more intelligent brain too had to wait for successive random genetic improvements, each improvement being followed by costly eliminations of all individuals that lacked it; selection was literally against stupidity rather than directly for intelligence.)

Two series of pre-adaptations probably prepared the way for evolution of a fine-grained, delayed-action, thinking brain. In one series the change from claw climbing to hand climbing, then to brachiating, and then from trees to the ground permitted successive increments in size of animals and of their brains, the larger brains having more space for complex recording and integration of information. In the other series selection in the arboreal environment favored finer perception, precise manipulation, socialization, and increased vocal communication, all of which favored brains able to integrate and respond to more finely divided factors in the environment. And socialization may have had another, decisive effect: social, cooperative acts need not be as rapid as self-preserving and competitive ones. Animals cannot stop to think when predators spring or competitors race for the same edible insect, but some cooperative and altruistic behaviors allow a time lapse in which the brain can compare complex alternatives. Of course this *is* speculative, and an oversimplification, but there probably was a partial shift during pre-human evolution from relatively simple, largely instantaneous behavioral repertories to increasingly complex, more delayed-action repertories that allow more time for trial and error, learning, and thinking.

Brain capabilities and resulting behaviors do not fossilize, but the sizes and sometimes the gross structures of brains can be calculated from the sizes and shapes of fossil brain cases, and a general relation between brain size and brain capability (in closely related animals) can reasonably be assumed. The brain increased in size during the arboreal stage of primate evolution—primates have relatively larger brains than most mammals—and fossils indicate that the increase continued rather slowly in the australopithecines, more rapidly when *Homo* diverged from *Australopithecus,* and explosively during the final evolution of *Homo sapiens.* Most of the increase was due to disproportionate increase in size and complexity of the cerebrum and especially of the cerebral cortex, which is concerned in human intelligence. The "bilaterality" of the brain—the different specializations of the two hemispheres—is unique to man and probably dates from the last, language-connected stage of human evolution; human individuals lack the bilaterality of the brain at birth, but it "emerges in the course of growth, coinciding with the development of speech . . . "(Campbell, page 355). The increase in rate of brain evolution in early *Homo* and the explosive increase of brain size and complexity in *sapiens* invite explanation.

The Brain and Throwing. If successive step-by-step adaptations, each itself selectively advantageous, were pre-adaptations that led to evolution of a fine-grained, delayed-action, thinking brain in man, why did not the same process lead to the same result in other primates? The answer probably is that something unique initiated a uniquely rapid period of brain evolution in the pre-human lineage. Wilson (*Sociobiology,* pages 568–569), following Jolly, suggests that the "prime impetus" may have been feeding on grass seeds, which increased hand dexterity, which pre-adapted the hands to using tools, but I doubt both the implied degree of specialization on grass-seed feeding and the suggested effect of it. A more logical, and more effective "prime impetus" to the evolution of man's brain may have been the evolution of effective throwing.

Modern evolutionists concentrate on what they can see and measure. This is good, but only up to a point. Some components of evolution that cannot yet be measured have probably been important too, and throwing may be one of them.

Anatomical modifications of the arm for throwing are not conspicuous. Fossil arm bones are rare. And thrown stones, unlike "pebble tools," were presumably scattered and not brought together at camp sites, where they can be recognized. There is, therefore, no clearly visible evidence of the evolution of man's throwing arm and throwing behavior, and students of human evolution have usually ignored it. Nevertheless, the human arm has *evolved* as an efficient, unique sling. No other animal can throw as man does. Other primates do throw sticks and stones, but only awkwardly. Van Lawick-Goodall, cited by Wilson (*Sociobiology*, page 173), records 44 objects thrown (tossed describes it better) by wild chimpanzees, with only 5 hits, all within 2 meters, and none damaging. Compare this with the force and especially the precision of human throwing: a skillful man has a good chance to break the skull of another man with one stone at 30 meters. This skill is evidently the result of the coordinated evolution of a bone-and-muscle throwing apparatus, a hand that can first hold and then release a thrown object effectively, and a brain that can program the complex act of throwing. Effective throwing may also have required an erect, well-balanced posture, and may have contributed to the evolution of the latter.

Although there is no direct evidence, there is some collateral evidence of the time available for the evolution of the complex physical and neurological mechanism of throwing. Robinson (1972, page 245) says that the posture, proportions, and "static and dynamic balance" of *H.* (= *Australopithecus*) *africanus* probably gave it "considerable capacity to use the body for such actions as wielding tools and weapons, throwing, and so forth." Australopithecines of the *africanus* type are now dated, as fossils, back 5 or 6 million years ago, and they or early *Homo* were making and using broken stone tools (which is more complex behavior than throwing unworked stones) at least $2\frac{1}{2}$ million years ago. There are therefore several million years available for the evolution of man's throwing arm and effective throwing behavior.

Throwing may have played three roles in human evolution. First, it was and sometimes still is directly useful in food getting, defense, and fighting. Throwing of spears and clubs is still used for all these purposes by Australian Aborigines, who do not have the bow. We ourselves still throw stones and clubs, if not often to kill game, at least to knock down fruit and nuts. We still defend ourselves by throwing or threatening to throw stones; dogs learn quickly that a man with a stone is dangerous, and earlier carnivores probably learned it quickly too. And "stoning" was still a means of fighting and execution in Biblical times and, more rarely, still is. All these direct uses of throwing probably began far back in pre-human history. A naturalist trying to photograph our ancestors several million years ago might have been kept at telephoto distance by thrown stones. These primary advantages of effective throwing may have started its evolution, but other effects of throwing may have been more important.

Second, throwing may have had decisive social effects. The thrown stone may have been the first effective "equalizer" (compare the six-gun in the Old West), which shifted selective advantage from brute force to skill (from Goliath to David) and from force-dominated hierarchies to more cooperative and diversified organizations.

And third, and most important, throwing may have played a special, decisive role in evolution of the human brain. It is a commonplace of evolution that structure and function evolve together. When, during the evolution of ape into man, the brain increased in size and complexity, it was presumably performing new functions. To say that its principal new function was exercise of intelligence is not enough. Intelligence is an imprecise concept, and intelligent behavior involves delay in responses and may be less advantageous to most animals than instantaneous stereotyped behavior. The evolving pre-human brain probably performed more specific functions, of more direct selective advantage. Effective throwing surely was selectively advantageous, and it probably required evolution of a brain with new, special characteristics.

Most animal behavior is concerned with not more than two principal moving objects (the animal itself and another animal), and the brain need not model the relationship between the two with great precision. But effective throwing requires *precise* correlation of *three* moving objects (the animal, the object thrown, and another animal). Probably nothing that apes do requires such complex correlation or such precision. The solving of three-body problems—the modeling of complex geometric patterns with the precision required for effective throwing—may have been a decisive factor in the evolution of the human brain. And the effectiveness of throwing was probably increased when the brain had evolved to the point at which it could criticize and correct throwing errors. When we pick up a handful of stones and throw them in succession, correcting toward a target, we may be doing one of the things that began to raise the pre-human brain above the ape's level (Figure 9).

Obviously, throwing was only one of many factors in the evolution of the human brain. It may have been directly important only during the early *Homo* stage of pre-human evolution, during which the brain increased moderately in size and presumably in capability. But during this stage throwing may have been a decisive factor in pre-adapting the

Figure 9. Speculative reconstruction of an early *Homo* soon after the split from *Australopithecus*, perhaps 3 or 4 million years ago, showing hair covering already reduced, posture already erect, preparation to throw with force and precision, and additional stones in hand to be thrown with corrections if the first throw misses.

brain to perform still more complex and precise functions later: in counting (the first step toward arithmetic and mathematics), geometry, and language. Of course this is only the bare outline of one hypothetical, invisible and unquantifiable—and therefore speculative—sequence of adaptation, pre-adaptation, and further adaptation in the complex evolution of man.

We now know that throwing ability varies and is at least partly heritable (Kolakowski and Malina, 1974), which is consistent with its evolution by selection. And the pleasure we take in throwing is significant too. Compare it with carrying. The habit of carrying food and, later, other useful objects in the hands must have been selectively advantageous for millions of years, but we do not take much pleasure in it; the desirability of the objects carried is enough of an inducement; and to carry useless objects in the hands for fun would *not* be advantageous. But carrying is a relatively simple behavior that hardly has to be learned at all; throwing, a much more complex behavior that must be learned and is improved by practice. It was therefore selectively advantageous for evolution to reinforce throwing by making it fun for our ape-man ancestors, and this is why it is still fun.

This hypothesis can be extended. Evolution evidently did reinforce precise throwing behavior by making it pleasurable, and the pleasure may have been extended to doing other complex activities precisely, that is, to taking pride and pleasure in good workmanship. This is another unquantifiable characteristic of man that presumably has an evolutionary origin. Of course this idea too is highly speculative!

The Brain and Language. The neurological mechanism of language—how the brain puts words together to express complex ideas—is still debated, but something can be said in general about it.

Chimpanzees communicate by facial expressions, behaviors, touch, and wordless sounds (as we still do too), but they do not have a vocal apparatus like ours and cannot pronounce words as we do. They can be taught a simple sign language and can put signs together into simple phrases, but what they can do barely suggests the beginning of human language. What a chimpanzee can do is evidently limited by its brain, which is inherently incapable of putting words together complexly.

Our pre-human ancestors presumably once had a communication system and brain comparable to the chimpanzees' (but I do *not* mean that they invented a hand-sign language, which chimpanzees learn only from man), and we can speculate about the later evolution of language in relation to brain evolution.

Language was probably used first to exchange increasingly complex information for immediate purposes—to signal not just "look" but look in front or behind, up or down, near or far, and at the speaker, or food, or a snake—and then to transmit information with delay from hour to hour, day to day, and generation to generation. The enormous selective advantage of this was that it allowed the experience of many individuals to be used by single individuals in making decisions and doing things. (This is comparable to the enormous advantage that a population gains by putting together the genetic improvements made in different demes; see Chapter 6B.) Single individuals can learn much for themselves during their lifetimes: where to find food, how to avoid pain, and so on. And they can learn by imitation. But if language allows them to use also what has been learned by other individuals out of sight, in the same and earlier generations, the additional advantage must be very great. Two heads are indeed better than one, and larger sets of brains are better still. The beginning of effective use of language may well have touched off the final, explosively rapid evolution of a uniquely large, complex, thinking brain in *Homo sapiens.*

Language and its evolution are under intense study now. Two (among many) fascinating but somewhat technical treatments are cited in the chapter reading and reference list.

Reinforcement and Esthetics. The evolution by selection of brain-determined behaviors is often "reinforced." Behavioral reinforcement occurs when an act is performed because it gives pleasure to the performer; pleasure is the reinforcing factor. Evolutionary reinforcement occurs when, during evolution, the frequency or effectiveness and therefore the selective force of an advantageous behavior is increased by making it pleasant to perform. Evolutionary reinforcement is obvious in sexual behavior and is visible or inferred in many other advantageous animal and human behaviors that are performed more often or more effectively because of it. Pleasure in present performance may be evidence that a behavior was strongly advantageous in the past. And reinforcement-by-pleasure of one behavior may have secondary effects that explain others that are otherwise difficult to understand. We can see this most clearly in ourselves because we can feel our own pleasure. We cannot directly perceive the pleasure of other animals, although we can measure its effects in simple experiments and can infer its broader effects.

Evolutionary reinforcement probably plays an important role in "behavioral ecology." Ecologists ought to ask, but do not often do so: why do animals stay in the environments they are adapted to? The animals surely do not calculate the advantages. The answer probably is that animals find their environments satisfying or pleasant. This seems to be the only answer that explains environmental choice by birds and mammals. Of course the satisfactions and pleasures involved in evolutionary reinforcement are not immaterial but have physiological, hormonal, or neurological bases, and can reasonably be considered reinforcing components of behavioral wholes, genetically determined, and capable of heritable variation and evolution by selection.

Reinforcement has probably been important in human evolution and may be the origin of many human emotions, including the pleasure men take in performing altruistic acts, in joining groups, and in throwing, hunting, and fighting. It may also be the origin of our esthetic emotions.

Our esthetic appreciations are primarily emotional. We do not enjoy a landscape or the sound of music because our reason tells us to, although our complex brains have greatly increased the complexity of the pictures and music we enjoy. Some esthetic emotions are evidently based on sex. But the pleasure we feel in looking at an appropriate piece of the environment, or a picture of it, may be the same sort of pleasure that an ape or a mouse feels in its environment and that keeps it there. And the pleasure we take in hearing the human voice and (by extension) music may have its evolutionary source in emotional reinforcement of social communication. Increasingly complex vocal communication was surely selectively advantageous during human evolution, and its selection was presumably reinforced and strengthened by making both the uttering and the hearing of appropriate sounds a pleasure.

Evolution of Brain and Intelligence: Summary. Beginning with and continuing the evolution of the primitive mammalian brain, which was a product of perhaps 500 million years of evolution, the primate brain evolved increasing size, complexity, and capability in response to (i. e., by selection in) arboreal environments. The increase continued with increase of social behavior, which was itself first a response to the arboreal environment but which became more complex and important when, after descent to the ground and spread to open country, organized hunting began. And use of tools, possibly beginning millions of years before they appear in the actual record, may have contributed to the continuing evolution of a more complex and capable brain.

However, within this continuum of brain evolution, two events had special consequences. The evolution of effective throwing—a selectively advantageous behavior that required a *complex precision* not required of any other pre-human activity—induced evolution of a brain pre-adapted to precise counting, arithmetic, geometry, and language; the (hypothetical) consequence was the moderate acceleration of brain evolution during the divergence and early evolution of *Homo*. And evolution of effective language—increasingly complex, precise, *delayable* communication, enormously advantageous because it allowed single individuals to use the experience of many—touched off the explosive evolution of *Homo sapiens*.

Evolutionary reinforcements of selectively advantageous behaviors—by adding pleasure to them—occurred in connection with choice of habitats, sexual behavior, social behavior, tool use, and (notably) effective throwing and vocal communication, and had secondary effects that may be the sources of our pleasure in good workmanship and of our esthetic appreciations.

This summary of evolution of the human brain and brain-determined behaviors combines evidence, inference, and speculation. It is consistent within itself and (I think) satisfying as far as it goes. But it is probably not all true and certainly not the whole truth. I offer it to naturalists only as a frame to be corrected and to fit details to.

Cultural Heredity and Evolution

Evolution of the human brain and language was accompanied by evolution of a form of heredity—cultural transmission—which, although it presumably began far back in pre-human history, has assumed a new, overwhelming importance in man, in whom cultural evolution now overlies and largely hides continuing genetic evolution. Because this subject is emotionally and ethically explosive, consideration of it will be postponed to Chapter 10.

Evolution of Human Societies and Civilizations

The evolution of individuals—their bodies, brains, and behaviors—has been correlated with the evolution of social behaviors, social groups, and group technologies of man's ancestors and man. A very rough and oversimplified continuity can be traced from nonsocial mammalian reproductive behavior; to (beginning possibly something like 50 million years ago) relatively simple primate social groups, with females and offspring but not males forming incipient families; to (beginning possibly less than 10 million years ago), stable families of male, female, and offspring, with increasing opportunities for division of labor and teaching; to (beginning possibly less than 6 million years ago) groups of several families of nomadic hunter-gatherers, held together by the advantages of organized hunting (and other advantages), with further opportunities for division of labor and teaching of increasingly diverse and specialized techniques and behaviors, with (later) the use of fire and (still later) the beginning of language and a rapid increase of cultural heredity; to (beginning only about 10,000 years ago) larger, permanent communities depending on farming and animal domestication, with increasing populations, increasingly complex technologies of tool use, transportation, communication, and social organization; to (within historic times) an explosively accelerating series of increasingly complex civilizations using successively bronze, iron and other metals, coal, and now atomic splitting and fusion.

This evolutionary continuity has not only itself accelerated explosively but also has produced an explosive increase of numbers of people, from a world population of possibly 10 million hunter-gatherers 20,000 years ago, to 4000 million people now, a 400 fold increase.

Of course this continuity has in fact been exceedingly complex. The stages have not simply followed each other but have formed broad transitions, and have overlapped, forming a cumulative sequence or stepped whole. I myself have seen, on the Bismarck Range in New Guinea during World War II, New-Stone-Age people who made beautifully finished stone axes but had never seen metal, and within memory some Australian Aborigines still used Old-Stone-Age stone-chipping or flaking techniques. And each stage has surely been in itself far more complex than we know.

Continuing Evolution in Man

Naturalists who look thoughtfully at what is going on around them will see that man is still evolving, by variation, transmission of variations, and selection-by-elimination, and that evolution is occurring at both genetic and cultural levels, with interactions between the two.

The genetic process is being studied intensely. A good introduction to human genetics is the "outline" by Edwards (1978); a highly regarded older text of human genetics, by Stern (1973); and a review of five more-recent texts, by Glass (1977). Close to 2000 gene-determined variations in man have been detected or suspected, and this is probably a small fraction of all that exist. The genes concerned range from simple Mendelian alleles that affect the color of hair or eyes, to complex sets of alleles like those involved in Rh blood incompatibility, to not-yet-identified genes suspected of predisposing individuals to schizophrenia. Natural selection in man is treated in several additional recent books, of which the best for naturalists is probably the one edited by Bajema (1971), which begins with a 15-page article by Dobzhansky on "Man and Natural Selection." Much of the genetic selection in human populations is not directional but stabilizing—elimination of new damaging mutants including those induced by radiation—but directional selection is probably continuing too. Its genetic base is more difficult to detect, but selective elimination of individuals with inferior backs, or feet, or hearts, or brains probably does continue processes of adaptation to our erect posture, increased activity, and complex cultures that began far back in pre-human history.

Continuing cultural evolution in human populations is more obvious, but even more difficult to analyze. However, it does involve variation, transmission, and selection-by-elimination, as noted in Chapter 10. Some of it is stabilizing—elimination of new inferior ideas and behaviors —but cultural evolution is a conspicuous directional process too.

The Cost of Man's Evolution

That genetic adaptive evolution imposes a heavy cost in elimination of enormous numbers of individuals is emphasized in Chapter 6B; the cost of evolving from a four-footed (or four-handed) to an erect, two-legged posture is suggested on a previous page; and the probably enormous cost of rapid evolution of a brain capable of human language and intelligence has been noted too. These few details (out of the great number that could be cited) are repeated to emphasize that man's body and brain are still far from perfectly adapted to man's environment, and that a heavy cost is still being paid, although (as suggested elsewhere) the cost of not evolving might have been still heavier.

Cultural evolution of new ideas, new behaviors, and new technologies costs relatively little immediately. But it remains to be seen (if anyone remains to see it) whether technological evolution may not impose a delayed cost payable in elimination of whole populations or of all mankind (see again Chapter 10).

Summary

Man is the inconceivably complex product of the whole of organic evolution. A narrow, usually one-species-wide ribbon carried by an unbroken succession of living individuals, which naturalists can visualize, has woven itself through 3 billion years of life, 1 billion years of multicellular eucaryotes, $\frac{1}{2}$ billion of vertebrates, 100 million of mammals, 50 million of primates, and shorter periods of apes, ape men, and ancestral man. Of course this list is drastically abridged and the times are rounded approximations. During these times the ribbon had added—woven into itself—characteristics of each successive group, including step-by-step pre-adaptations (see Chapter 6C) to the making of man. And now, at the unfinished end of the still evolving ribbon, man combines characteristics derived from all the successive groups in his history with notable and unique characteristics of his own.

The last stages of man's evolution have made him an eye-oriented, fine-discriminating, dextrous, inquisitive, imitative primate; notably vocal, sexy, social, selfish-altruistic, and dominating-submissive; uniquely naked, erect, two-legged, with hands free to carry, throw, and manipulate; with a long attention span, a unique vocal apparatus, and a unique, fine-grained, delayed-action brain capable of complex language, delayed communication, and thinking; and with a unique capacity for cultural heredity and rapid evolution of new behaviors, social organizations, technologies, and philosophies. The cost of man's evolution has been heavy, and is still being paid. But the cost of not evolving would probably have been heavier still—extinction long ago.

READING AND REFERENCES

The deluge of publications on human evolution must bewilder conscientious naturalist-readers. Cited here can be only a few general books that are widely available, and a few more-technical works on subjects that are critical (for example, the fossil record of man's immediate ancestors), controversial (for example, the hunting hypothesis), or previously slighted (for example, throwing in relation to the evolution of man). The sequence of headings in the list is therefore even more abridged than in the other chapters. For further explanation of my rather informal reading and reference lists, see the Preface and Chapter 3.

General Reading

Campbell, B. G. 1974. *Human Evolution. An Introduction to Man's Adaptations,* 2nd ed. Chicago, Illinois, Aldine. (Probably the best full-scale, readable treatment of the subject.)

Wood, B. A. 1978. *Human Evolution.* London, England, Chapman and Hall; New York, Halstead (Wiley). (A concise, up-to-date summary.)

Wilson, E. O. 1975. *Sociobiology. The New Synthesis.* Cambridge, Massachusetts, Harvard Univ. Press. (Brings together much information and speculation on the evolution of societies and social behaviors of animals, including man. *See also* my Chapter 10.)

Leakey, R. E., and R. Lewin. 1977. *Origins. What New Discoveries Reveal about the Emergence of Our Species and Its Possible Future.* London, England, Macdonald and Janes;

New York, Dutton. (Includes a well-told account of new fossil evidence, but prejudiced against the hunting hypothesis. *See also* my Chapter 10.)

Coon, C. S. 1954. *The Story of Man*. New York, Knopf. (Outdated, but still good reading.)

Van Lawick-Goodall, J. 1971. *In the Shadow of Man*. Boston, Massachusetts, Houghton Mifflin. (An outstanding book on wild chimpanzees, with evolutionary implications.)

Sagan, C. 1977. *The Dragons of Eden. Speculations on the Evolution of Human Intelligence*. New York, Random House; paperback, Ballantine, 1978; London, England, Hodder and Stoughton, 1978. [Exciting reading and widely read, but serious naturalists should read it to test their speculative power, not for information. The review in *Nature* (Vol. **272**, 768–769) says of it, "... inaccurate, full of fanciful and unilluminating analogies, infuriatingly unsystematic, and ... unerringly concentrating on the superficial and misleading at the expense of the important and significant. That it should have had such success is a sad commentary on the power of publicity."]

SECTION A

Under a Magic Spotlight

Romer, A. S. 1968. *The Procession of Life*. (*See* Chapter 8.)

Vertebrate Continuities

Romer, A. S. 1968. (See earlier citation.)

Miles. A. E. W. 1972. *Teeth and Their Origins*, Oxford Biology Readers, No. 21. (*See* citations in Chapter 2.)

Weale, R. A. 1978. *The Vertebrate Eye*, 2nd ed. Oxford Carolina Biology Readers, No. 71. (*See* preceding citation.)

SECTION B

The Naked Ape

Baker, M. A. 1979. A Brain-Cooling System in Mammals. *Scientific American*, **240**, May, 130–139.

SECTION C

The Fossil Record: Man Apes to Man

Pilbeam, D., and S. J. Gould. 1974. Size and Scaling in Human Evolution. *Science*, **186**, 892–901.

Leakey, R. E., and R. Lewin. 1977. (See under "General reading.")

Gould, S. J. 1979. Our Greatest Evolutionary Step. *Natural History,* **88**, June–July, 40–44.

Human Races and Their Evolution

Coon, C. S. 1962. *The Origin of Races.* New York, Knopf. (A controversial. history, criticized in the present book.)

Darlington, P. J., Jr. 1957. *Zoogeography: The Geographical Distribution of Animals.* New York. Wiley; London, England, Chapman and Hall. (The last chapter is concerned with the zoogeography and dispersal of man and is pertinent to the origin of races.)

The Hunting Hypothesis

Read, C. 1925. *The Origin of Man.* Cambridge, England, Cambridge Univ. Press. (Includes an early statement of the hunting hypothesis.)

SECTION D

Evolution of Brain and Intelligence

Smith, C. U. M. 1970. *The Brain, Towards an Understanding.* London, England, Faber and Faber. (A book I have found especially readable and informative.)

Various authors. 1979. *The Brain. Scientific American,* September. (A set of articles on the brain and how it works.)

The Brain and Throwing

Darlington, P. J., Jr. 1975. Group Selection, Altruism, Reinforcement, and Throwing in Human Evolution. *Proc. Nat. Acad. Sci. USA,* **72**, 3748–3752.

Robinson, J. T. 1972. *Early Hominid Posture and Locomotion.* Chicago, Illinois, Univ. Chicago Press.

Kolakowski, D., and R. M. Malina. 1974. Spatial Ability, Throwing Accuracy and Man's Hunting Heritage. *Nature,* **251**, 410–412.

The Brain and Language

Lieberman, P. 1975. *On the Origins of Language. An Introduction to the Evolution of Human Speech.* New York, Macmillan.

Rumbaugh, D. M. (ed.). 1977. *Language Learning by a Chimpanzee. The Lana Project.* New York, Academic.

Reinforcement and Esthetics

Darlington, P. J., Jr. 1975 (*See* earlier citation under "The brain and throwing.")

Evolution of Human Societies and Civilizations

Darlington, C. D. 1969. *The Evolution of Man and Society.* London, England, George Allen & Unwin. [A 400,000-word survey by a distinguished geneticist-evolutionist who is also a generalist and intellectual maverick of "how man and his societies came into being." Reviewers have called the book controversial and infuriating, and also well reasoned, meriting close study, to be read and digested at leisure. The review in *Nature* (Vol. **224**, page 193) indicates the wide scope of the book; the one in *Time* (October 17, 1969, pages 74–75) cites some of its provocative ideas. Serious naturalists should read the book at their leisure, critically, and will find a 30-page list of references for further reading, classified by subjects.]

Continuing Evolution in Man

Edwards, J. H. 1978. *Human Genetics.* London, England, Chapman and Hall; New York, Halstead (Wiley).

Stern, C. 1973. *The Principles of Human Genetics,* 3rd ed. San Francisco, California; Freeman.

Glass, B., 1977. [Review of five books on human genetics.] *Quart. Rev. Biol.,* **52,** 409–410.

Bajema, C. J. (ed.). 1971. *Natural Selection in Human Populations. The Measurement of Ongoing Genetic Evolution in Contemporary Societies.* New York, Wiley.

PART FOUR

Evolutionary Philosophy and Ethics

Something so huge that it seemed unfair to man that it should move so softly stalked splendidly by them . . .

Lord Dunsany, *The Book of Wonder* (quoted from the Boston edition, page 30); not said of evolution, but surely applicable to it.

Nothing in nature is absurd, though much is deplorable.

Rex Stout, *A Right to Die*, page 5 (put into the mouth of Nero Wolfe).

10

PHILOSOPHICAL AND ETHICAL
IMPLICATIONS OF EVOLUTION

I am neither a philosopher nor a moralist and shall not try to write as either but as a naturalist and evolutionist with some special knowledge, writing about everyday matters important to common people, for common people to think about. By common people I mean ordinary, intelligent men and women with some common sense. Of course I shall stress what naturalists can see in the world around them and in themselves, and how what they see may help them understand human problems.

Principles

Although this chapter is concerned chiefly with aspects of philosophy and ethics that are related to evolution, some broader principles require attention first. I shall state them simply and may oversimplify them. But I suspect that essential principles of philosophy may be as simple as essential principles of evolution are, and that philosophers as well as evolutionists may sometimes lose sight of the simple principles in their excitement over abstruse complexities.

Explanation. Primitive folklore, modern religions, philosophies, and scientific theories including the theory of evolution form a spectrum of beliefs and hypotheses which have common characteristics. They are all primarily explanations, and secondarily they allow predictions (often mistaken), a hope of some control of future events (also often mistaken), and rules of conduct. The need for explanation seems to be inherent in man. A reasonable hypothesis is that inquisitiveness was selectively advantageous to our prehuman ancestors because it led to discovery of new foods and detection of dangers—it must have led some individuals into danger too, but it presumably conferred a net advantage—and that the need for explanations evolved from it as the human brain evolved, the habit of explaining becoming itself selectively advantageous because it led to better ways of doing, making, and using things. A related hypothesis is that stereotyped behaviors that reduced competition among individuals and favored evolution of reciprocal altruism were advantageous to our ancestors, and that the stereotyped behaviors evolved into rules of conduct. The systems of explanation and the rules of conduct probably evolved together, the explanations generating the rules, and selection at both individual and group levels setting limits to both. But the mode of evolution probably itself evolved during this history, from mainly genetic in man's pre-human ancestors to mainly cultural in man.

Scientists do not lose the need for explanation but only change their explanations when they have to, as Darwin did when he exchanged the Biblical explanation of the origin of living things for an evolutionary explanation.

The Reductionist Frame.   Scientific explanations differ from others in being based on a great range of verifiable observations. Scientific explanations therefore have a kind of reality that other explanations do not have. But their reality is confined within an assumption, which is that the "real world" that scientists see can be reduced (or could be if we had the omniscience to do it) to a wholly material system which obeys physical and chemical laws. All scientific work is done within this "reductionist frame"; explanations that go outside it are not scientific. Some scientists try to look outside the reductionist frame, and this is legitimate if they make clear the point at which they stop talking as scientists and start talking as science fictionists or part-time mystics, and some other scientists repudiate reduction without specifying an alternative, which seems pointless if not craven.

Truth: What Is It?   "What is truth? said jesting Pilate; and would not stay for an answer" (Francis Bacon). And the answer does not come quickly. Truth, like beauty, is in the eye or the mind of the beholder. In scientists' minds truth is what scientists observe in the "real world," but there are limits to it. They can observe only within the reductionist frame. And while scientific *observations* can be verified, explanations cannot be, but can only be tested and "falsified," or not falsified, endlessly. This is the hypothetico-deductive process referred to in Chapter 3. Scientific explanations therefore evolve toward final truth but never reach it. But scientific explanations can become true in the everyday meaning of the word, as I think the theory of evolution has.

A separate question, considered at the end of this chapter, is: how much of the truth as he sees it should an evolutionist tell?

Mysteries

Science does not deny mysteries but tries to solve them.

Complexity versus Mysticism.   The human brain is an infinitesimal part of reality, and the models it makes *must* be absurdly oversimplified. We should accept this fact, and should then be able to accept the inconceivable complexity of reality. I do it comfortably. I accept that I cannot understand, for example, infinity, or the origin of matter, or why we are here at all, without the need of mystic explanations. And I do not see that changing from a religious to a pragmatic point of view (when I was about 12 years old) and to a reductionist-evolutionist one later has resulted in my psychological or ethical deterioration. Our ethical systems of behavior depend on our altruistic emotions, which are real, rather than on mystic philosophies or religions. That religious people are not necessarily more ethical than other people is a commonplace of history and novelists.

Holism and Emergent Evolution.   In the past, but less often now, some evolutionists have turned to *holism* (Smuts)-the doctrine that an organized whole is more than the sum of its parts—and emergent evolution (Wheeler)—the doctrine that unpredictable new characteristics of wholes "emerge" during evolution. These doctrines have often carried mystic implications. However, although a whole is indeed more than the sum of its separate parts, reductionists think that it is not more than the sum of the parts and of their interaction (Chapter 4). If we cannot predict the characteristics of a new whole, this is because we do not know enough to do it. Holism and emergent evolution are only parts of a continuing and, I think, continually unsuccessful effort of some evolutionists to reconcile evolution, mystic ideas, and religions. Teilhard de Chardin was perhaps the best-known evolutionist who was still trying to reconcile them until his death a few years ago—after which the Vatican condemned his works.

The Mystery of Matter.   I do not mean that there are no mysteries left in science. The outstanding mystery now seems to me to be the origin of matter—why matter exists (has evolved?) capable of forming itself into the most complex patterns of life. It is a fundamental evolutionary generalization that no external agent imposes life on matter. Matter takes the forms it does because it has the inherent capacity to do so. Selection then eliminates some and permits survival of other forms, in suitable environments and for limited times. This is one of the most remarkable and mysterious facts about our universe: that matter exists that has the capacity to form itself into the most complex patterns of life. By this I do *not* mean to suggest the existence of a vital force or entelechy or universal intelligence, but just to state an attribute of matter as represented by the atoms and molecules we know. Given this attribute, it was inevitable that matter would evolve to living levels on the earth's surface, and it is also inevitable that acting sets will evolve to comparable levels of organized complexity wherever matter like ours exists in suitable environments, although the paths and products of evolution are sure (within the statistical limits of sureness) to be very different in different places. But all this is a mystery because we cannot conceive of all the factors and all the complexity of it. We do not solve the mystery by using our inadequate brains to invent mystic explanations.

Evolutionary Considerations

The evolution of man is treated in Chapter 9. Now to be considered are special aspects of evolution that concern our philosophies and ethics.

Organic evolution is like an enormous, multi-level engine, described in the summary of Chapter 6. Two levels of it particularly concern man.

At one level is the main Darwinian process, which operates by excess reproduction, genetic variation, and selective elimination of great numbers of individuals. Evolution at this level has made existing organisms in all their diversity; it has made our bodies and brains and the parts of our emotions and behaviors that are genetically determined; and it still operates in man as well as in other organisms.

Cultural Heredity and Evolution   At another, higher level (higher because it is superimposed on the other) a new process, a sort of overdrive, has been added. It affects, not our genetically determined characteristics, but behaviors and ideas that are not determined in detail by our genes.

Beginnings of this process can be seen among animals that imitate each other. For example, in England a few individual birds have discovered how to pick the tops off milk bottles and get the cream, and the discovery has been handed on, by imitation, to other birds of the same and following (but overlapping) generations. An example nearer our own ancestry is the behavior of some chimpanzees of breaking a hole into a nest of termites, pushing in a twig, and pulling it out and eating the termites clinging to it (see van Lawick-Goodall in Chapter 9). This was probably invented by accident (supplemented by intelligence?) by one lucky chimp, who was imitated by associates and offspring. Our actual ancestors probably transmitted many behaviors by imitation before the evolution of language, including perhaps the making and using of gradually improved stone tools.

The evolution of language greatly amplified cultural heredity and cultural evolution in man. But this obvious fact must be qualified in two ways. First, imitation is still important in man, and it does not always lead in the same direction as intelligent teaching. How often do parents say to their children, "Do as I say, not as I do," only to see the children imitate the parents' actions rather than heed their words? And second, genes can still change human behaviors; medical research has found many genes that limit or change what our brains as well as our bodies do. We should therefore try to look at our behavior objectively, and try to distinguish the genetic and nongenetic components in it.

The evolution of a genetically determined behavior must wait for random gene mutations and recombinations to occur, and then must wait for the reproductive cells bearing the genes to develop into individuals before the behaviors can be expressed, and then requires the selective elimination—premature deaths or reproductive failures—of enormous numbers of individuals. But cultural evolution is relatively rapid, and cheap. It requires only formulation of new ideas and new combinations of ideas, which may be virtually instantaneous in their origin and expression, and which can then evolve by selective elimination of ideas rather than of individuals, although individuals may be eliminated too if their ideas lead them in dangerous directions.

Because they are relatively rapid and cheap, cultural processes might be expected to run far ahead of genetic processes in evolution of behavior in man. And we may speculate that the two processes compete. Imagine an advantageous new behavior evolving in two competing populations, in one genetically and in the other culturally. Its evolution would probably be more rapid in the second population, which would eliminate the first, and a secondary effect might be that the new cultural process would eliminate the old genetic mode of evolution. To the arguments against detailed genetic determinism in, for example, kin selection (Chapter 7) can be added the speculation that genetic determinism of human behavior may have been reduced because evolution has found a better way.

And speculation can be carried further. Cultural evolution involves something like inheritance of acquired characters. This fact is noted in Chapter 5, with Prichard's comment that if "all acquired conditions of the body" did not end with the lives of the individuals in whom they are produced, the universe would be filled with "monstrous shapes." Acquired conditions of the mind do not all end with the lives of the individuals in whom they are produced. They are directly transmitted from mind to mind, and they evolve and diversify in the process. And some persons might say that the universe of man's ideas *is* therefore filled with monstrous shapes—fashions that run their uncontrolled courses in human behavior, philosophy, art, and even science. Extreme examples are the mystic cults we see proliferating within our own culture: my impression is that human minds still run off in uncontrolled directions as far and as fast as they ever did, and that this may threaten our survival as much as atomic technologies do. The same thing occurs in science: most new ideas in science are probably wrong; many are falsified in the minds that produce them before they are transmitted to other minds; but some are transmitted and flourish fashionably for a while without control. In all these cases the "monstrous shapes" evolve rapidly and cheaply, not limited by selection in each generation as genetic evolution is. But costs may be deferred rather than avoided. In the end the monstrous shapes may disintegrate almost of their own monstrosity, and the cost may then be catastrophic: the elimination of whole ways of life, philosophies, esthetic standards, and scientific theories to which many generations have contributed, and sometimes the elimination of whole populations whose cultures turn out to be noncompetitive or self-destructive. But although most new ideas come to nothing or end disastrously, some do lead to breakthroughs. The compulsion to pursue ideas without too much immediate control may contribute to effective cultural evolution. ("Control" in this connection means self-control by intelligent elimination of unrealistic ideas. Social control of ideas, whether by reactionary politicians or activist biologists, would be disastrous, and should be tolerated only when the ideas are *directly,* violently destructive.)

But these are speculations, not conclusions. In trying to assess the roles of genetic and cultural processes in man, we should admit that theory leaves the conclusion open, and we should *look at people,* at the ordinary people around us, to see, or try to see, what they really do and why they do it.

Sociobiology and Determinism. Edward O. Wilson's *Sociobiology* (1975) applies biological and evolutionary principles to the study of societies and social behaviors. The book was received with admiration—and Wilson's treatment of nonhuman societies is admirable—but then touched off a furious controversy about "determinism" in man. Our genes do make us human. The controversy is about whether genes determine our behavior in detail, or whether they only broadly pre-adapt us to evolve diverse behaviors by cultural rather than genetic processes.

Genetic determinism does not require that one gene determines one behavior, which is performed if the gene is present and not if it is absent. Genes may be present in all individuals of a population but may induce behaviors only in special situations, but the genes may then determine the behaviors precisely. See the tri-determinant model of an ant colony, under "Social groups" in Chapter 7. And compare how genes act during the ontogeny of a multicellular organism: all genes are present in all cells, but many of them act only in special situations, although they then determine the cell's behaviors precisely. So, we should not reject determinism simply because we do not find simple Mendelian relationships between genes and behaviors. We should *look* at real people to try to see how they do behave. Both sides of the controversy should be looked at critically.

Determinists emphasize kin selection (my Chapter 7), and say or imply that human altruism is precisely, genetically adjusted to degrees of kinship, to the profits of selfish cheating, and to the costs of countering it. But when we look at real people we see that, although man is indeed an altruistic animal, his altruism is *not* precise in detail and is *not* precisely limited. Our altruism often extends beyond the limits of man to cats and whales and even plants. Moreover, simple selectively advantageous behaviors such as odor-oriented ones and instinctive ability to swim have *not* evolved as strict determinist theory would predict. I think that critics who try to be objective will probably decide that strict determinists make unjustified assumptions based on simplistic mathematics (Chapter 3) and on failure to look at real people.

Antideterminists are represented by the signers of the letter that began the controversy, in the *New York Review of Books*, November 13, 1975. The letter is headed "Against Sociobiology," and is signed by 16 persons including competent evolutionists. It asserts that "strong scientific arguments . . . show the absurdity of these [determinist] theories . . ." But this assertion is itself absurd; the theories in question are not absurd, only probably wrong. The letter then says that Wilson's book "conceals political assumptions" and is "a defense of the status quo" supported by "strategies and sleights of hand." (These are only samples of the exaggerations and unjustified assertions in the letter.) This seems to accuse Wilson of motives that are, to say the least, unscientific. But I know Wilson well, and I know that he tries to tell the truth honestly, although (like all of us!) he may make mistakes. Objective readers should ask whether persons who use the vocabulary and strategies of this letter to try to discredit honest science do not by their own words discredit themselves and convict themselves of "political assumptions." (As to the assertion made by some antisociobiologists that all science is subjective, see under "Truth: how much?" later in this chapter.) Critics still pour out diatribes against sociobiology, but most of them do little more than repeat the assertions in this letter. The repetition may be good propaganda, but it is not good science.

How can uncommitted persons—naturalists—decide between the possibly simplistic determinists and the possibly political antideterminists? They can try to *look* objectively at the evidence. Then they will see, I think, not only that our genes make us human, but also that genes from our outmoded past probably still influence our behavior in a general way, although not precisely. Take, as an example, the "hunting hypothesis." Antideterminists assure untutored readers that there is "no evidence" for it, but this is another absurd assertion. "Evidence" includes "indications," indirect as well as direct, and there are many indications (listed in Chapter 9) that group hunting influenced our evolution and still influences our behavior, although not in such precise detail as some determinists suggest. Human altruism too looks gene-based but not precisely gene-determined (as just noted). And human intelligence, discussed next, is surely gene-based and looks as if it varied genetically, but without being *precisely* gene-determined.

These examples suggest what might be called a practical, general (as opposed to a theoretical, precise) determinism. They suggest that man does inherit general characteristics, determined by selection in the past, that (in Nero Wolfe's words) are deplorable but not absurd, and that can be ignored only at our peril.

Human Variation and Intelligence

Human beings are genetically diverse. We easily distinguish by visible details all the individual men and women we meet, except identical twins, and we see what seems to be great variation in intelligence too.

An emotion-loaded question now is, what is the relative importance of genetic and environmental factors in intelligence? An evolutionist's answer is that our brain and our intelligence have evolved together; that their evolution from ape to man has surely been primarily genetic, which is to say by genetic variation followed by selective elimination of less intelligent and survival of more intelligent individuals; that the genetic variation of brain and intelligence that was essential to our evolution probably still occurs at least in part; but that the genetic determinants are often greatly modified by the environment. This is, I think, both good evolutionary logic and consistent with what we see.

We should, therefore, accept that *both* genetic *and* environmental factors *may* affect human intelligence and behavior, and we should make practical decisions accordingly. Schools should *both* allow individuals who may differ genetically in intelligence to proceed at different rates or by different paths, *and* provide help for environmentally handicapped individuals. And society should allow males and females—who do differ genetically, and who must play different primary roles in reproduction—*either* to play conventional, secondary, family roles if they wish, *or* to have equal opportunity to play other roles. These are common-sense decisions, already made by many people—examples of the common sense of common people.

Race. Subspecies or races—distinguishable, usually geographic subpopulations—occur in many species (Chapter 6C). They are products of the continuing evolution of local populations that goes on everywhere and always and that is one of the strongest evidences of the fact of evolution (Chapter 2).

The races of man are briefly described in Chapter 9. Anthropologists argue bitterly about the classification and evolution of human races, the arguments sometimes being heated by cries of "Racist!" and counter-cries of "Defamation!" But these arguments are out of order here. I want here only to say what human races are as seen by an evolutionist who is not a racist—a "racist" being defined in dictionaries as someone who believes that some races are inherently "superior," and who discriminates against "inferior" ones, which I do not believe and do not do —and to say something about the ethical and practical problems that human races pose.

Races of man share the complexities of races of other organisms. Each human race is complex within itself. Each includes an enormous range of individual diversity, which is partly genetic, partly environmental or cultural. Human races (like those of other organisms) are complexly interrelated and poorly bounded. And all human races are interfertile; the gene pools of all of them share most of the same genes; the genes that actually distinguish races—that determines skin color, for example—are few and irregularly distributed. (Genes that distinguish related species may be few too; distinct species of *Drosophila* may differ in only a few genes; and even man and the chimpanzee apparently differ in only a small fraction of their genes.)

The prehistoric evolution of human races is considered briefly in Chapter 9. At the dawn of history the principal races occupied fairly well-defined geographic regions, but even then their boundaries were indistinct and changing. Some races overlapped and intergraded. Some were spreading and others were retreating, and both replacements and mixings occurred continually.

For example, according to Coon (1954, pages 319–320), certain food plants including the taro were brought to East Africa from India by unknown traders long ago, and the techniques of iron smelting and making iron weapons were introduced later. The better-fed and better-armed Negroes of eastern Africa then multiplied, became dominant, and spread at the expense of other races. Southward-spreading Negroes reached South Africa about the time of White settlement, and replaced most of the earlier, sparser populations of Bushmen there. Most of the Negroes in South Africa eventually outdistanced or abandoned the taro and primitive iron smelting and adopted new foods and new technologies brought to South Africa by other races. All this culminated in the present racial crisis in South Africa.

For another example, before civilization reached the southern tip of South America, three groups of people, corresponding to three ethnic groups, lived on the big island of Tierra del Fuego. They spoke languages so different that they could not understand each other. One, on the outermost tip of land, was on the point of extinction, and the other two occupied ecologically different zones, one the interior forest and low mountains and the other the channels and islands along the western edge of land, and the two groups competed murderously where they came together. (This is from an extraordinary book by E. L. Bridges, *The Uttermost Part of the Earth*. He was a missionary's son, almost the first European born on Tierra del Fuego, who spent part of his early life living with the Indians. He gives a firsthand account, as I think no other book on any part of the world does, of the impact of Europeans on a primitive culture from first contact to final deterioration. The book is not rare and is rewarding reading for naturalists.)

The process of spreading of some races and extinction or mixing of others is still going on, conspicuously in Amazonian Brazil and less conspicuously in many other parts of the world.

Because races are so complex, because they differ so little genetically, because they overlap and intergrade, and because of fear of "racist" interpretations, some biologists deny that human races exist at all, but I think this is mistaken, and useless. What is the use of telling common people in suburbs of Boston or London or in South Africa that there are no races, when common people can see the differences? Would it reduce emotions and the (often economic) causes of emotions if we taught people to say "You damned multivariate" instead of "You damned ————"? For biologists to deny the reality of race, I think, simply marks them as fools or sophists in the eyes of common people, and weakens their influence in other matters in which they should be listened to. It is both more honest and more useful to say what races are in fact, and to admit that they have always generated prejudice and conflict, as they still do. Then, recognizing the fact rather than trying unsuccessfully to hide it, we may hope to find modes of education and government that reduce racial tensions and make mixed communities more ethical.

Ethics

Ethics involve value judgments—judgments of what is good or bad— that are not scientific. But knowledge of evolutionary principles and of our own evolutionary history can help us make judgments that are effective because they recognize inherent human limits, conflicts, and needs. Evolutionists may therefore legitimately write about ethics, as I am doing. Remember that I am writing as a common man who is also a naturalist and evolutionist, writing about everyday matters important to common people, for common people to think about. If some of my conclusions are trite, this is because they are (I hope) based on common sense, and common people have a lot of common sense which has already led them to the same conclusions. But an evolutionary point of view may clarify even some trite conclusions, and may suggest some that are not trite.

Individuals and Populations. There is an inherent conflict of interest between the individual and the population it belongs to. Evolution has produced individuals that must do *both* what is good for them as individuals *and* what is good for their populations, or that must *both* seek food when hungry, escape danger if they can, and fight if they must to stay alive, *and* expend energy and run risks to reproduce. And evolution has produced a complex system of emotional compulsions, punishments, and rewards that further these conflicting interests. This conflict of interest is ineradicably built into us and is at the root of some of our most intractable human problems. But most evolutionists either ignore this conflict or persuade themselves that it does not exist. One difficulty is that biologists usually think of anything that increases the representation of an individual's genes in the next generation as good for the individual, but this is not what the individual human thinks, and it is not a basis for a workable system of ethics. A workable system for man must combine selfishness and altruism—must reconcile satisfactions for individuals with the conflicting requirements of other individuals in the population and with provisions for future generations. Such a system requires continual compromises that are complex and difficult.

Selfishness and Altruism.   Although the conflict can be thought of as between the individual and the population, it is actually between the individual and other individuals in the population. Individuals both compete and cooperate, and have both selfish and altruistic relationships.

Selfishness is an essential characteristic of genes and individuals (as noted in Chapter 7), and has been since the origin of life. It has been maintained by selection. Genes and individuals that did not take care of themselves long enough to reproduce (or to gain representation in the next generation in some other way) removed themselves from evolutionary sequences, leaving evolution to their selfish competitors. Selfishness is inherent in man because evolution has made us selfish.

Unselfishness too is inherent in genes and individuals, which must at least contribute to the success of their derivatives (copies of themselves, or their offspring) if they are to have a place in evolution, and which to do this cooperate with other genes and individuals. This involves altruism (see again Chapter 7), which is always (or almost always) potentially reciprocal and profitable. Altruists are statistically more likely to gain than to lose by their altruism, and this is true of human individuals no less than of individual ants in an ant colony, or of the fungus and alga in a lichen. But humans do not calculate the profits of altruism, or usually do not do so. They behave altruistically because altruistic behavior is innate, generated by compulsions and reinforced by the pleasure altruistic acts give the altruists. It is indeed often more pleasant, as well as more blessed, to give altruistically than to receive.

Effective ethical systems are based on man's innate, uncalculated altruism but must recognize and control man's innate selfishness too; most of the Ten Commandments are concerned with this.

Authority and Antiauthority.   Ethical systems depend on acceptance of authority by most individuals most of the time. The authority may be mystic, vested in one class of persons or a single person with supposedly divine knowledge or power; or maintained by force, by one class or one dictator; or vested in a group as a whole, by common consent. It is, I think, not pushing speculation too far to suggest that ethical systems that require *both* exercise of authority *and* acceptance of it have evolved from dominance hierarchies that began among our remotest social ancestors and that were perhaps strengthened during the evolution of hunting groups (Chapter 9). But all ethical systems are, or may be, opposed by antiauthority. They may be weakened or destroyed by selfish evasion and cheating, or modified and strengthened (or sometimes destroyed) by unselfish dissent and revolt.

"Cheaters" (Chapter 7) are individuals who accept altruistic benefits but do not return them. They are conspicuous in most human populations and are countered in various ways. If they are not countered, they impose increasing loads on populations, and if the loads become too great, the populations may be destroyed. "Cheaters" are inherently destructive. The only good that might be said of them is that they may promote evolution of intelligence by imposing on and eliminating unintelligent individuals, but they do this "good" at high cost to populations.

Dissent and revolt are or may be unselfish (although the motives of the individuals concerned may be mixed). They introduce new ideas into ethical systems, promote diversity, and promote evolution of ideas. They are comparable to mutations, recombinations, and selective processes in the evolution of genetic systems. They therefore play useful, perhaps essential roles in the evolution of ethical systems. But too much dissent and revolt can be destructive too, just as too many mutations can.

Aggression and Violence.   Aggression and violence are painfully in the public eye now, and much written about. They may be either selfish and destructive (violent crime) or unselfish and idealistic at least in intent (some, but not all, political violence). I shall not try to analyze them in detail, but shall make some points about them as an evolutionist sees them.

The first point is that selfishness and violence are inherent in us, inherited from our remotest animal ancestors. They are not peculiar to man. Some biologists have tried to persuade themselves that cooperation rather than competition is the rule in nature, and that violence (a form of competition) is unnatural or secondary, but they are mistaken. Nature is, conspicuously, red in tooth and claw, and I do not see how naturalists who look carefully at the world around them can doubt it. It is one of the things that Rex Stout probably had in mind when he made

Nero Wolfe say that nothing in nature is absurd although much is deplorable. Violence is, then, natural to man, a product of evolution.

The second point is that, although inherent in man, most human individuals do not practice violence for its own sake but only in response to fear, or pain, or needs, although the needs may be psychological as well as material. To the extent that violence is a response, it should be possible to reduce it by reducing the stimuli that evoke it.

The third point is that violence is indeed deplorable now—outmoded —in human communities; that we would like to reduce it; and that to do so we must understand it. Most books on violence are concerned with the roots of it, with why we are violent. Of course this is important, but study of its reduction would be more so. Violence is conspicuous in history, including our own history. How and why has it been reduced in some communities, including ours in the United States? And in spite of the recent increase of violent crime, we are less violent than our ancestors were or than many other communities are now. We need only read the Bible, or early English history, or current foreign news in the daily papers to see that this is conspicuously true. So, I suggest as most needed now, not more study of violence, but a critical study of nonviolence, or of how violence can be and sometimes has been reduced.

The fourth point about violence is that communities must counter it, with counter violence if necessary, if they are to survive. Counterviolence is deplorable too, but may sometimes be necessary to survival of the social-ethical systems that we prefer to their alternatives.

And finally, a question that is unfashionable among humanitarians must be asked and answered in a practical world: what happens to a community that refuses to use violence to defend itself against a violent competitor?

The Rights of Man.   As a biologist and evolutionist, I cannot see that "rights" exist among animals and plants in the real world. Individual plants and animals have no "right" to share resources, live, or reproduce, but do so in the face of competition (sometimes reduced by cooperation) as best they can with the means evolution has given them. When individual animals establish territories, they are not held by right, but by force, or threat of force, although the threats may become conventional. The same situation holds, more complexly, in man. The idea of "natural rights of man," meaning of human individuals, seems both not realistic and not useful. It seems better from both points of view— reality and usefulness—to think of human rights as conventions that are desired and agreed on by most individuals in a community, and that are maintained by the individuals and the community by whatever means are necessary and available. Rights of this sort do exist and do form ethical patterns.

Why "All Men Are Created Equal." Obviously all men are not equal biologically or socially. Different individuals are very unequally advantaged or disadvantaged, genetically and in the environments they are born into. The concept of equality of rights, especially equality of opportunity, is the basis of our society, but the rights are not inborn but are established by agreement among us and maintained, or not maintained, by our actions. This kind of society and the kind of government it generates—democracy—is very far from perfect. Winston Churchill called democracy "the worst kind of government—excepting all the others." But given human beings as they are, even our imperfect approximation of it is still the best kind, the kind that is best for most individuals, and the kind that is most susceptible to improvement by common consent. "All men are created equal" is, then, not a biological fact but an ethical decision—that all men and women shall have equal rights and opportunities—made and supported by common people.

Social Darwinism. "Social Darwinism" is the application of Darwin's theory of natural selection to human situations. It has been supposed to justify "survival of the fittest" (a phrase invented by Herbert Spencer, not Darwin) and "the devil [unfettered competition] take the hindmost" especially in industry. It has therefore been heavily condemned by philosophers writing about ethics, and blamed for some of the worst effects especially of the industrial revolution. But the industrial revolution began in the century before Darwin, and was disastrous for many individuals, including many children, long before publication of the *Origin.* Darwinism has been unfairly blamed for it, but was not the cause; this distinction between attempted blame and actual cause is essential to real understanding. Man's inhumanity to man is in fact worldwide and as old as history, and it still continues, more or less, everywhere, including places where Darwin has never been heard of. To blame him for it is dangerous rationalization, dangerous because it hides the real cause, which lies in inherent human nature.

Genetic Engineering. Gene-splicing (DNA recombination), cloning, and so on (these techniques need not be described here) are being studied intensely in microorganisms and may become possible in man, although probably not soon. They open the possibility of increasing control of human evolution, and they therefore require ethical decisions. First, a decision must be made about whether they can safely be permitted in organisms other than man. This decision is relatively simple and has already been reached by common-sense majorities: the probable benefits are greater than the dangers, and the techniques should be permitted, with adequate safeguards, and without forgetting that human error (and occasionally human evasion) may be more dangerous than technical miscalculation. And second, a decision will have to be made about when and how to apply the techniques to man. This decision, too, should be made, when the time comes, by common-sense majorities of common people, after full information and discussion.

Extinction and Conservation Ethics

Extinction is a natural process essential to evolution, as noted in Chapter 6C. Now to be considered are man's role in it, and ethical implications.

This is a difficult subject for me to write about. Many conservationists whom I respect, including good friends of mine, will not like what I say. But the subject is evolution-related, and I have to treat it.

Man's evolution, multiplication, and occupation of the world have inevitably caused the extinction of many plants and animals, directly or indirectly. Man has hunted or is hunting many animals to extinction, either for food (for example, the Dodo on Mauritius, some of the giant tortoises on the Galápagos, and probably the moas in New Zealand), for sport (for example, the Ostrich in Arabia), or in self-defense (for example, the Lion, which has been retreating before man for 2000 years). Current lists of extinct and vanishing species include many more examples. But it has been "man's role in changing the face of the earth" that has caused the most massive extinctions.

Man's impact has been greatest on islands like those just named, but may sometimes have been decisive on continents too. The spread of man through North America late in the Pleistocene may have caused the extinction of many large mammals, including mammoths and giant sloths. But details are doubtful. In some parts of the world, including South America and Australia, many large Pliocene and Pleistocene mammals became extinct before the arrival of man. And many large mammals survived the evolution and initial impact of man in Africa.

Moreover, some of even the most recent extinctions would perhaps have occurred in any case. The Passenger Pigeon nested and migrated in enormous flocks that were presumably vulnerable to epidemics of disease and to local catastrophes, which might have destroyed it. And the Carolina Paroquet, which disappeared about the same time, was the only parrot widely distributed in the north-temperate zone, which is marginal for parrots, and from which all other parrots (known fossil in Europe in the Tertiary) had become extinct long before the arrival of man.

Some persons assert that all species have an equal "right" to existence, but this dubious argument is not likely to persuade many persons to become conservationists. And it is neither necessarily true nor very persuasive to argue that we should preserve species because species diversity promotes ecological stability; the relation between diversity and stability is in fact complex and doubtful (Chapter 6C).

However, there are other, stronger, practical arguments in favor of conservation based on intelligent understanding of particular situations. There are strong arguments in favor of conserving forested watersheds, wetlands, "green belts," extensive natural recreation areas, productive forests, and wilderness. And beyond this it is probably wise to conserve substantial samples of as many different natural communities as possible, in their full diversity, as potential sources of useful species (for example, fungi that produce antibiotics) or of genetic materials for improvement of crop plants and domestic animals (by hybridization and perhaps by "genetic engineering").

Within this context of intelligent and useful conservation, there is room for reasonable efforts to save individual species that are scientifically interesting or that give pleasure to naturalists. But I do not think we should try to save every local form of every species, regardless of cost. We should recognize that saving an almost-unique paddlefish is more important than saving one out of many species of darters. And we should be forthright. If we oppose a dam because it threatens wilderness, we should fight it on that ground and not because it might drown out a lousewort. And if we do make the lousewort an issue, we—concerned naturalists—should take the trouble to find out what the plant's range really is, and whether the dam would flood it completely. Above all, we should look at conservation problems as wholes. We should not try to block a dam in Maine, atomic power in New Hampshire, strip mining in the West, and a pipeline across Alaska as if they were unrelated. A dam in Maine would cost some wilderness, but water power is clean, and its use reduces demand for power from atoms, coal, and oil. I do not mean that I advocate the dam in Maine, but only that it should be judged as part of a nationwide or worldwide problem. Conservationists have too often lost credibility and done their cause real harm—weakened it in the eyes of common people—by poor judgment, exaggeration of isolated cases, and unintentional misrepresentation. Public protest is useful; it focuses attention on important problems; but it should be used to emphasize the problems, not to force decisions.

Books uncritically advocating conservation are probably in every public library now. Suggested to bring balance to reading on the subject is Thomas (1975), *The Follies of Conservation,* which "shows how easily a genuine concern can become so over-sentimentalized that the tide of opinion aroused may, through shortsightedness, unwittingly work against the very cause it set out to support."

Truth: How Much?

Evolutionists (like other scientists) must ask themselves not only, What is truth? (a question considered earlier in this chapter), but also, how much of the truth as I see it should I tell? The reflex answer is, the whole truth, and the argument for it is that even deplorable truth is better faced than evaded. But after making this answer, I ask myself, are there exceptions to it? And immediately I see one: in comparing human races it seems best, at least for the time being, to suspend judgment, not to try to decide whether (for example) one race is more intelligent than another. I do not mean that the truth should be suppressed or distorted in this case, but that investigation of it is best postponed, because at present investigation cannot be separated from emotion, and because it too often leads to half-truths which are used to justify discrimination, violence, and genocide. But a counterargument can be offered here. Although scientists' conclusions may (wrongly) be used to try to *justify* discrimination, they do not *cause* it. This is another case (as in "Social Darwinism") in which attempted blame and actual cause should be carefully distinguished. Racial discrimination, violence, and genocide have always been with us, as again the Bible and history show. They did not wait for (supposed) scientific justification, and do not require it now, but are products of emotions that are very old, but now outmoded.

Some scientists assert that all science, including the study of evolution, is so strongly colored by social beliefs and prejudices that there can be no such thing as objective truth. This statement is usually made by scientists whose ideas do seem to be prejudiced or "political," and it is reminiscent of Aesop's fox who lost his tail. Those who have lost their objectivity would like us to believe that we all have lost it, or ought to lose it. But most of us *try* for objective truth in both observation and explanation, and if some scientists do not, their science is the worse for it—less useful and more dangerous than it ought to be.

Most evolutionists will, I think, agree that we should almost always try to tell the whole truth as we see it, as objectively as possible. This is what has made science so effective and useful to man.

Politics, Propaganda, Suppression, and Activism. Politics, propaganda, and suppression of ideas—selfish use and abuse of truth—ought to have no place in science, but in fact they do. The fashion now is to treat the use of truth as almost more important than truth itself.

Politics is defined (in my Webster's Collegiate Dictionary) in several ways, including "the art or science concerned with winning and holding control over a government" and "political activities distinguished by artful and often dishonest practices." To mistake other persons' motives is one of the most damaging mistakes that people make. I shall try not to do it, and shall say only that some biologists and evolutionists act as *if* they calculated their science to win control over university departments and over presentation of science to the public, and as *if* they calculated artful and dishonest methods to do it.

By "dishonest methods" I mean telling part truths calculated to persuade rather than to inform, or deliberately telling people, not the whole truth or (when the truth is unknown) not what all the reasonable alternatives are, but calculated parts of the truth, or of what the teller thinks or hopes is true. It amounts to telling people, not what is true, but what the teller thinks is good for them, or good for the teller. I do not know whether this distortion of truth is deliberate or not—probably it sometimes is and sometimes is not—but it is misrepresentation. It is dangerous in single cases and much more dangerous as a fashion.

Propaganda is a weapon of politics in science as well as in other human activities. It may be used in good faith, as I think promathematical propaganda and counterpropaganda are (Chapter 3), or it many be used "for the purpose of helping or injuring an institution, a cause, or a person" (again from Webster's *Collegiate Dictionary*). It is sometimes difficult to distinguish between legitimate publicity of a scientific finding and exaggerated or distorted (propagandist) claims, and between fair criticism and suppression of ideas, especially since criticism and "falsification" of ideas are an essential part of science.

But outright suppression is not falsification and is out of place. It occurs when ideas are countered, not in open discussion, but by keeping them from being discussed at all, and by suppressing or punishing individuals who try to discuss them. Examples range from suppression of dissent in religion, sometimes by violent means (the Inquisition), and suppression of new scientific ideas by threat of punishment (Galileo), to efforts to suppress teaching of evolution in schools, and efforts by "activists" to block scientific discussion of (for example) sociobiology.

Activists are (I think) essential in political communities and science. They are analogous to mutants in genetic evolution; they cause diversification and contribute to the evolution of ideas. But two comments on this analogy are in order. Most genetic mutants prove not useful and have to be eliminated, at high cost (Chapter 6B), and most activist ideas too are likely to prove not useful and will have to be eliminated, sometimes at high cost. And outstandingly novel ideas in evolution have in fact usually come, not from what we now call activists, but from individuals less confident of their own ideas and less in a hurry to push them: from Darwin rather than Samuel Butler, and from Mendel rather than (say) Haldane.

Nevertheless, activists are, I think, essential in science. But there are several categories of them. How can common people, including myself, judge them? We might begin by distinguishing constructive activists, who try to make sure that what they think (sometimes mistakenly) are good ideas will be heard by scientists and governments, and destructive activists, who try to prevent other persons' ideas from being heard. In general, and put too simply, constructive activists are useful, destructive, not.

Both kinds of activists use a wide range of methods which require further judgments. Even constructive activists, anxious to insure that what they think are correct decisions will be made, sometimes use force or violence. This is usually bad in itself, and even if the immediate results are good, use of force and violence is shortsighted. Communities must and do counter them by greater force and, if necessary, violence, which then too easily become tools of government accepted by most common people. I accept them, reluctantly, as and if activists make them necessary.

As to negative activists who use force and violence to interrupt scientific meetings and silence opposing speakers, I see nothing in their favor. I see no essential difference between them and the persecutors of Galileo, or Hitler's enforcers. A writer in *Science* argues (in effect) that suppression is permissible if used against ordinary scientists; he asks, "Who do they [the persecuted scientists] think they are, Galileos?" Surely common people can see what is wrong with this question! And common people will, I think, see the contradiction when destructive activists, after themselves using force, complain when force is used to restrain them.

If these conclusions are trite—as I warned they would be—it is because they are common-sense conclusions already reached by common people.

Three further comments may be useful. First, activists sometimes seem to want to suppress ideas—for example, the idea that violence is inherent in us, in part inherited from hunting ancestors—because they think the ideas are dangerous. But surely it is better to face danger and try to counter it than to turn one's back on it; even the jackal in Kipling's *The Undertakers* knew this. Second, misunderstanding other person's motives is one of the commonest causes of trouble among human beings; it is a principal reason why older and younger generations do not understand each other; and it is conspicuous in arguments between conventional and activist scientists. And third, common people can at least try to evaluate arguments among scientists not only by the merits of the cases but also by the way the cases are presented. When, for example, activists say that there is "no evidence" for the hunting hypothesis, although in fact there are many reasons for considering it seriously (Chapter 9); or when they say that sociobiologists are motivated by desire for publicity and self-aggrandizement, although in fact the ones I know try to tell the truth objectively; or when, on the other side, sociobiologists overemphasize activists' admitted prejudices and ignore their legitimate arguments, common people will do well to discount what both sides say.

Summary: The Naturalist's Role

In closing I want to return to a theme that is introduced in the preface and runs through the whole of this book: the role of naturalists in the study of evolution and in the application of evolutionary concepts to man.

Good naturalists will see—or remember if they cannot see—not only that a whole is more than the sum of its parts (but not necessarily more than the sum of the parts and their inconceivably complex interactions) but also that a part within a whole is less than its separate self; the whole modifies and limits what a part does (Chapter 4).

Good naturalists will see—or try to see—the living world as a whole. Professional biologists, who are usually specialists, often see only parts of it, and often do not see how the parts fit into the whole. Good naturalists will monitor and criticize what biologists say always looking at the real world, including real people, to do it. Of course naturalists should try to understand as well as criticize what biologists say, and will see the real world more clearly if they do.

Specifically, good naturalists will see that much evolutionary mathemathics is unrealistic because it treats parts rather than wholes (Chapter 3) and will not mistake mathematical models for reality, but will look to see if the mathematical predictions are fulfilled. Often they are not. It should be part of the role of good naturalists to guard against wrong decisions in human affairs based on unrealistic mathematical models.

Good naturalists will see that, within the multilevel whole, evolution at the Darwinian level—the visible level of individuals and populations— has made us what we are, and in making us, has made also our appalling problems, which rise from population explosions and from "human nature" as it has evolved from our pre-human ancestors, and they will see that this is the level that must be understood if the problems are to be solved. New discoveries at molecular levels are exciting and illuminating, but their human importance is sometimes grossly exaggerated.

From reading this book, good naturalists should understand the simple principles of Darwinian evolution (stated in the "Summary of summaries" at the end of Chapter 6), and should remember that in natural populations excess reproduction condemns enormous numbers of individuals to premature death, which sentimental conservationists sometimes forget, and that the cost of selection (Chapter 6B) limits what evolution can do, so that we should not expect perfection either in the plants and animals we see or in ourselves.

Good naturalists will look critically at both determinist and antideterminist positions (Chapter 10), and will (I think) see that the truth is somewhere in between. Good naturalists will see that complex races of man exist, and that selfishness, aggression, and violence (but also altruism) are inherent in man as well as in many other animals. But they will see also that it is safer to face than to try to hide deplorable realities, and that the realities do not forbid equality of opportunity and human rights, agreed on and maintained by common consent.

Good naturalists will discount the well-meaning exaggerations and propaganda of some conservationists, will try to see situations as wholes, and will work for intelligent choices in conservation rather than simplistic solutions.

Wholes should be looked at too in making practical decisions in everyday matters such as budgets, taxes, and legal systems. A system of 100 laws may be intolerable as a whole even if each law is good in itself; legislators and administrators sometimes do not see this, but good naturalists should.

Good naturalists will see that most scientists—an inarticulate majority —*try* to tell the truth objectively and will distrust the few—a vocal minority—who try to color, hide, or suppress truth. Good naturalists will not only defend truth, but will defend also the scientists who try to tell it.

And finally, good naturalists will (I hope) be able comfortably to accept the unknowable, inconceivable complexity of reality without recourse to mysticism.

READING AND REFERENCES

This can be only a minute sample of the great number of recent books on various aspects of evolutionary philosophy and ethics, but some books will lead to others, and library subject indices should add more.

Principles

The Reductionist Frame

Ayala, F. J., and T. Dobzhansky (eds.). 1974. *Studies in the Philosophy of Biology.* London, England, Macmillan.

Mysteries

Holism and Emergent Evolution

Smuts, J. C. 1926. *Holism and Evolution.* London, England, Macmillan; Compass Books ed., 1961, New York, Viking.

Wheeler, W. M. 1928. *Emergent Evolution and the Development of Societies.* New York, Norton.

Teilhard de Chardin, P. 1959. *The Phenomenon of Man* (transl. of 1955 French ed.). London, England, Collins; revised English ed. 1965, New York, Harper & Row.

Evolutionary Considerations

Cultural Heredity and Evolution

Van Lawick-Goodall, J. 1971. (*See* Chapter 9.)

Sociobiology and Determinism

Wilson, E. O. 1975. *Sociobiology. The New Synthesis.* (*See* Chapter 9.)

Human Variation and Intelligence

Singer, S. 1978. *Human Genetics. An Introduction to the Principles of Heredity.* San Francisco, California, Freeman. (Short; well illustrated; with chapter reference lists.)

Osborne, R. T., C. E. Noble, N. Weyl (eds.). 1978. *Human Variation.* New York, Academic. (Largely concerned with variation of intelligence.)

Race

Brues, A. M. 1977. *People and Races*. New York and London, England, Macmillan.

Ebling, F. J. (ed.). 1975. *Racial Variation in Man*. New York, Halstead (Wiley).

Coon, C. S. 1954. (*See* my Chapter 9.)

Ethics

Aggression and Violence

Lorenz, K. 1966. *On Aggression* (transl. of 1963 German ed.) New York, Harcourt, Brace & World.

Russell, C., and W. M. S. Russell. 1968. *Violence, Monkeys and Man*. London, England, Macmillan. (In part an answer to Lorenz; see review by V. Reynolds, *Nature*, **221**, 99.)

Baron, R. A. 1977. *Human Aggression*. New York, Plenum. (The contents are indicated in *AAAS Science Books & Films*, **14**, 206.)

Genetic Engineering

Goodfield, J. 1977. *Playing God: Genetic Engineering and the Manipulation of Life*. New York, Random House; London, England, Hutchinson. (This and the following book are reviewed by David Baltimore, *Nature*, **272**, 766–767.)

Cooke, R. 1977. *Improving on Nature: The Brave New World of Genetic Engineering*. New York, Quadrangle/New York Times.

Extinction and Conservation Ethics

Thomas, W. L., Jr., et al. 1956. *Man's Role in Changing the Face of the Earth*. Chicago, Illinois, Univ. Chicago Press.

Mosimann, J. E. and P. S. Martin. 1975. Simulating Overkill by Paleoindians. Did Man Hunt the Giant Mammals of the New World to Extinction? *American Scientist*, **63**, 304–313.

Thomas, A. 1975. *The Follies of Conservation*. Elms Court, Devon, Arthur H. Stockwell Ltd.

INDEX

This index is designed for this book, which is a serious study of evolution but not a reference volume. It is essentially an index of subjects which readers will want to find or which I want them to find. Persons are indexed only if something significant is said about them. In the case of frequently occurring words such as "Darwin," "Evolution," "Man," and "Naturalist" only selected pages are indexed.